Lecture Notes in Mathematics

A collection of informal reports and seminars
Edited by A. Dold, Heidelberg and B. Eckmann, Zürich

T0255431

207

Leon Bernstein
Illinois Institute of Technology, Chicago, IL/USA

The Jacobi-Perron Algorithm Its Theory and Application

Springer-Verlag
Berlin · Heidelberg · New York 1971

AMS Subject Classifications (1970): 12 A 99

ISBN 3-540-05497-9 Springer-Verlag Berlin · Heidelberg · New York
ISBN 0-387-05497-9 Springer-Verlag New York · Heidelberg · Berlin

Offsetdruck: Julius Beltz, Hemsbach/Bergstr.

TABLE OF CONTENTS

Introduction

"Peut-être parviendra-t-on à déduire de là (namely - des circumstances remarquables, aux-quelles donne lieu la réduction des formes, dont les coefficiens dépendent des racines d'équations algébraiques à coefficiens entiers) un système complet de caractères pour chaque espèce de ce genre de quantités, analogue p.ex. à ceux, que donne la théorie des fractions continues pour le racines des équations du second degré" - with these words Charles Hermite, in one of his number-theoric letters [14] to C. G. J. Jacobi, challenged a great master of nineteenth century mathematics. The citation in brackets is due to P. Bachman [1]. The arithmetic characterization of quadratic irrationalities had been established by Lagrange's theorem stating that the expansion of a real number as a simple continued fraction is periodic if and only if it is the real root of a quadratic equation in one variable with rational coefficients. Since then mathematicians, in Dantesque despair, had abandoned all hope to expect any further information about the arithmetic properties of higher degree algebraic irrationals from the "naive" world of simple continued fractions. These, however, continued to attract the creative imagination of restless mathematicians who tried to generalize the Euclidean algorithm; these efforts lead, first and foremost, to the f-expansions of Bissinger [6], and other mathematicians like Everett, Good, Leighton, Renyi, Rogers and Pitchers [7] should be named among those who developed the theory of f-expansions and other binary generalizations.

C. G. J. Jacobi responded to Hermite's challenge with an ingenious sparkle; though his investigations of the subject were published only post-mortem in 1868 [15], he set out to work on the problem, in his own words, as early as 1839; at that time he already mastered his new tool of approximating real algebraic irrationalities of degree higher than 2, an instrument that has later acquired citizenship in our literature under the name of the Jacobi

Algorithm. It seems to the author that the delay in publishing his results, are definitely to be explained by the difficulties Jacobi encountered in treating the periodicity aspects of his algorithm. But let the great master speak by himself [15] about his most significant generalization of the binary expansion of real numbers to the dimension three:

"Es seien a, a_1, a_2 unbestimmte Zahlen, dagegen

$$\ell, m;\ \ell_1, m_1;\ \ell_2, m_2; \ldots$$

gegebene Grössen; ferner

$$a + \ell a_1 + m a_2 = a_3,$$
$$a_1 + \ell_1 a_2 + m_1 a_3 = a_4,$$
$$\cdots\cdots\cdots\cdots$$
$$a_i + \ell_i a_{i+1} + m_i a_{i+2} = a_{i+3},$$

so kann man durch successive Substitutionen sowohl a_{i+3}, a_{i+2}, a_{i+1} durch a, a_1, a_2 als auch umgekehet a, a_1, a_2 durch a_{i+1}, a_{i+2}, a_{i+3} ausdrücken..."

The Jacobi Algorithm then essentially proceeds in the following manner: starting with a triple of three real numbers u_0, v_0, w_0, new triples of numbers are generated by the recursion formulas

$$u_{i+1} = v_i - \ell_i u_i,$$
$$v_{i+1} = w_i - m_i u_i,$$
$$w_{i+1} = u_i,$$
$$(i = 0,1,\ldots;\ \ell_0 = \ell,\ m_0 = m).$$

This is the homogeneous form of the algorithm; in this book the non-homogeneous form is used, viz.

$$\frac{v_{i+1}}{u_{i+1}} = \frac{\dfrac{w_i}{u_i} - m_i}{\dfrac{v_i}{u_i} - \ell_i}, \quad \frac{w_{i+1}}{u_{i+1}} = \frac{1}{\dfrac{v_i}{u_i} - \ell_i},$$

with the notation

$$\frac{v_i}{u_i} = a_1^{(i)}, \quad \frac{w_i}{u_i} = a_2^{(i)}, \quad \ell_i = b_1^{(i)}, \quad m_i = b_2^{(i)},$$

$$(i = 0,1,\ldots;\ u_0 = 1)$$

so that Jacobi's Algorithm takes the form

$$a_1^{(i+1)} = \frac{a_2^{(i)} - b_2^{(i)}}{a_1^{(i)} - b_1^{(i)}}, \qquad a_2^{(i+1)} = \frac{1}{a_1^{(i)} - b_1^{(i)}},$$

and since, for binary expansions, one starts only with the initial values u_0; w_0, this takes the form

$$a^{(i+1)} = \frac{1}{a^{(i)} - b^{(i)}} \quad \text{or} \quad a^{(i)} = b^{(i)} + \frac{1}{a^{(i+1)}}$$

which is the Euclidean algorithm, if $b^{(i)} = [a^{(i)}]$. Indeed, for the calculation of his ℓ, m; ℓ_1, m_1; ℓ_2, m_2;... Jacobi used the same technique, viz. $b_1^{(i)} = [a_1^{(i)}]$, $b_2^{(i)} = [a_2^{(i)}]$, and it is shown in this book that it is exactly this point that caused Jacobi difficulties in search for periodicity of his algorithm. In order to express the initial values u_0, v_0, w_0 through their successor triples u_i, v_i, w_i, Jacobi needed his substitutions $a_i + \ell_i a_{i+1} + m_i a_{i+2} = a_{i+3}$; these little a's are being denoted in this book, following Oskar Perron [22], by $A_i^{(v)}$ $(i = 0,1,2;\ v = 0,1,\ldots)$ and already Jacobi has proved that the matrices $(A_i^{(v)})$ are unimodular. As to the initial values u_0, v_0, w_0 we hear again Jacobi saying in [15]:

> "Ich will im Folgenden für u_0 eine ganze Zahl und für v_0 und w_0 Ausdrücke von der Form
>
> $$a + bx + cx^2$$
>
> setzen, in welchen a, b, c ganze Zahlen sind, und x eine reelle Wurzel einer irreductiblen cubischen Gleichung bezeichnet, in welcher der Coeffizient von x^3 gleich 1 und die Coeffizienten der drei übrigen Glieder reelle ganze Zahlen sind."

Thus Jacobi chose v_0 and w_0 to be third degree algebraic integers; that the choice of the starting triple u_0, v_0, w_0 is a crucial point in the question of periodicity of the Jacobi algorithm, will emerge from the papers of various other authors and is also a fundamental issue treated in this monograph. But there is also another point that seems to have escaped the attention of Jacobi and those who followed his footsteps; they all took it for granted that the derivation of the $b_1^{(i)}$, $b_2^{(i)}$ from the $a_1^{(i)}$, $a_2^{(i)}$ must follow Euclid's example of putting $b_1^{(i)} = [a_1^{(i)}]$, $b_2^{(i)} = [a_2^{(i)}]$. That this is only one of the few ways (and rarely the best one) of calculating the $b_1^{(i)}$, $b_2^{(i)}$, is emphasized in this book over and again. This generalization of the Jacobi algorithm has one and only one urgently demanded aim in mind: that such an algorithm become periodic. For it is periodicity of the algorithm of a (finite) set of algebraic irrationals that answers Hermite's profound question about their arithmetic characterization; and the question then — still challengingly open — whether the Jacobi algorithm of any set of algebraic irrationals becomes periodic, is a matter of olympic curiosity only; to say nothing about a (finite) set of transcendental irrationalities — for it has been proved by Jacobi, later by P. H. Daus [9], D. N. Lehmer [18] for the cubic case, and then by O. Perron [22] and by this author in the general case $n \geqslant 3$, that if the Jacobi algorithm (or the generalized one) of $n - 1$ numbers ($n \geqslant 2$) becomes periodic, then these numbers all belong to an algebraic number field of degree $\leqslant n$. Here is what Jacobi says in [15] about his search for periodicity:

> "Um mich in diesen Algorithmen näher zu orientieren und zu sehen, ob die Quotienten ℓ_i and m_i, wie bei der Verwandlung einer Quadratzahl in einen Kettenbruch, periodisch wiederkehren, habe ich, als ich zuerst im Jahre 1839 diesen Gegenstand untersuchte, mehrere Beispiele berechnet, welche ich hier mittheilen will, da seitdem mehrere Mathematiker sich mit ähnlichen Untersuchungen zu beschäftigen angefangen haben, denen solche ziemlich mühsame Beispiele zur Aufstellung einer vollständigen Theorie zu Anhaltspunkten dienen können..."

Jacobi then demonstrates periodicity of the following triples by his algorithm:

$$1, \sqrt[3]{2}, \sqrt[3]{4};$$
$$1, \sqrt[3]{3}, \sqrt[3]{9};$$
$$1, \sqrt[3]{5}; \sqrt[3]{25}.$$

The reader will see later that Jacobi's examples are all special cases of infinite classes of periodic Jacobi algorithms; if, instead of the second example one uses the triple $1, \sqrt[3]{9}, \sqrt[3]{81} = 3\sqrt[3]{3}$, which is in the same field $Q(\sqrt[3]{3})$, then the first two examples belong to the class of triples

$$1, w, w^2, w = \sqrt[3]{D^3 + 1}; \quad (D = 1,2)$$

and the last example to the class

$$1, w, w^2; w = \sqrt[3]{D^3 - 3}; \quad (D = 2)$$

the expansion of these triples by the Jacobi algorithm (or its generalization) are stated explicitly in this book.

With the exception of a short paper by P. Bachman [1] who established necessary conditions for periodicity of the Jacobi algorithm in the cubic case, in form of restrictions on the $b_1^{(v)}$, $b_2^{(v)}$, literally nothing was done to promote the great idea of Jacobi until the end of the nineteenth and the beginning of the twentieth centuries; the years 1899 and 1907 mark a breakthrough in this subject and inaugurated a new epoch of deep investigations into the challenging problem of the Jacobi algorithm and its manifold applications. Let us start from the end: in that year the young Perron [22], then 27 years old - he became ninety on the seventh of May in this year 1970 and is living in Munich still working feverishly on the periodicity question of the Jacobi algorithm - published his dissertation (stretching over 76 pages) on this subject; his profound, most detailed and up-to-date analysis became the vade mecum of all young mathematicians working in this field. Introducing suitable notational symbols, Perron, first of all, raised the Jacobi algorithm from its isolationary cubism to the general n-th dimension, starting with an n-tuple of numbers $a_1^{(0)},\dots,a_n^{(0)}$ (in the original: $\alpha_1,\dots,\alpha_n$). One of his main merits is the study of the convergence of the algorithm, in analogy to the convergence of a continued fraction, but generalized to the n-th dimension, as will be explained in a later chapter of this book. That, in this sense, the Jacobi algorithm is always convergent, was masterly proved by Perron - for the first time, since Jacobi and his followers had silently (though correctly) presumed that this must always occur. But Perron went deeper; he asked the question: an infinite sequence of n-tuples $\left\langle b_1^{(v)},\dots,b_n^{(v)}\right\rangle$ given, does the limit of the ratio $\sum_{j=0}^{n} b_j^{(v)} A_i^{(v+j)} \Big/ \sum_{j=0}^{n} b_j^{(v)} A_0^{(v+j)}$ always exist; and he gave necessary conditions for this to happen. Much space of Perron's immortal paper is devoted to the characteristic equation (i.e. the equation of degree n + 1 whose roots generate the algebraic number field to which the initial n-tuple of numbers $a_1^{(0)},\dots,a_n^{(0)}$ belong). The crucial question of reducibility of the characteristic equation was thoroughly explored, especially in the case of a periodic algorithm. As the reader will soon learn from the following chapter, much can be read off from this characteristic equation of the algorithm. But fundamental as Perron's results may be, the young master did not succeed to answer Hermite's challenge in full and to give a complete arithmetic characterization of algebraic irrationals by the Jacobi algorithm. For this would mean: periodicity of the algorithm, yes or not. Now Perron proved "not" in case the components of the n-tuple $(a_1^{(0)},\dots,a_n^{(0)})$ are linearly dependent. But this would not satisfy Hermite; for one can always choose an n-tuple of linearly independent algebraic irrationals of a field of degree n + 1, and still nothing can be said about the "yes or not" of the periodicity of the Jacobi algorithm of this n-tuple. Closely connected with the periodicity of the Jacobi algorithm is also the question of best approximation of the $a_i^{(0)}$ by rational numbers - a situation we have been spoiled with by simple continued fractions, and which, however, does not carry over to the n-th dimension; for Perron has proved that if the Jacobi algorithm becomes periodic, the approximation law

$$\left| \frac{A_i^{(v)}}{A_0^{(v)}} - a_i^{(0)} \right| < \frac{C}{A_0^{(v)} \sqrt[n]{A_0^{(v)}}} ,$$

where C is a constant, independent of v, and n is the dimension of the initial triple of the algorithm, holds only for n = 1 (simple continued fractions) and for n = 2, whereby the conjugates of $a_1^{(0)}$ and $a_2^{(0)}$ are complex. In honor of the great master, the notation Jacobi-Perron algorithm was introduced by the author in previous papers.

The end of the nineteenth century was highlighted by the eruption of Minkowski's creative mathematical world: in 1899 he gave a complete answer to Hermite's question by supplying an exhaustive arithmetic characterization to real and complex algebraic irrationalities in any degree $n \geqslant 2$ in [20,a]. The method is based on a generalization of continued fractions by means of a geometric interpretation [20,b]; but his algorithm, also proceeding along linear substitutions, is not the Jacobi algorithm, though it always leads to periodicity, and that is its main and undisputable merit. Since Minkowski's method of substitutions is entirely different from Jacobi's, it is not displayed in this monograph. Also, its calculations are quite cumbersome; but this was true in Minkowski's days when arithmetic was "hand-made", not in the century of computers. With this in mind, the Jacobi algorithm may have lost some of its applicational value. That this is not so will hopefully emerge from the following chapters. The author also generalized the Jacobi algorithm in such a way that periodicity always ensues, and very soon - namely after n operations (n is the dimension of the initial triple).

The end of the nineteenth century also marks the publications of Woronoj [27] who proposed a new algorithm for cubic fields of both positive and negative discriminants, leading always to periodicity. Woronoj uses a minimal basis of the cubic field for his initial triple, and he also demonstrated how such a minimal basis can be effectively constructed. This periodic algorithm further leads to finding a fundamental unit in the cubic field, and that is the great merit of this new method, regardless of the most gigantic calculations involved. But Woronoj's algorithm does not generalize to algebraic fields of degree higher than three. An algorithm to calculate the fundamental units of cubic and biquadratic fields was recently published by K. K. Bilevich [5] who also claims that his method could be easily generalized to algebraic number fields of any degree $n \geqslant 5$. Applications of Bilevich's method to the calculation of fundamental units of biquadratic fields, carried out in a computer program by R. Finkelstein and H. London [12] do not seem, as I was informed by the authors, to justify Bilevich's claim of practicality for his algorithm, brilliant as his ideas are from a theoretical viewpoint.

The pioneering works of Jacobi, Minkowski and Woronoj have inspired almost all mathematicians of this century who again and again took up this difficult subject. Highlights of development are papers by W. E. H. Berwick [4], P. H. Daus [9], and George N. Raney [24]. But they all operate with modified Jacobi algorithms, as did Daus, or pursue ideas of Minkowski, as did Berwick, who links the expansion of an initial tuple of numbers in a cubic field with ideals of the field; but his substitutions are not even unimodular. New tools are introduced by G. N. Raney by using the n-dimensional unimodular group whose elements are the n x n matrices with integral rational entries and determinant 1, while Daus and Berwick restrict their investigations to the cubic case only. In this context also a paper by Joseph A. Raab [23] should be mentioned. An entirely new horizon in the study of the Jacobi algorithm was introduced recently by a series of most significant papers by Fritz Schweiger [25]. Pursuing ideas by Khintchine [16], S. Hartmann, E. Marczewski, C. Ryll-Nardzewski [13], N. Dunford and S. Miller [11], K. Knopp [17] and others, Schweiger develops an exhaustive theory concerning the metric and ergodic properties of the Jacobi algorithm. The tools of his deep investigations are the following. Let $x^{(0)} = (x_1^{(0)}, x_2^{(0)}, \ldots, x_n^{(0)})$ be a point in the n-dimensional unit cube E_n; a transformation of E_n into itself is given by

$$x^{(0)}T = x^{(1)} = \left(\frac{x_2^{(0)}}{x_1^{(0)}} - \left[\frac{x_2^{(0)}}{x_1^{(0)}}\right], \frac{x_3^{(0)}}{x_1^{(0)}} - \left[\frac{x_3^{(0)}}{x_1^{(0)}}\right], \ldots, \frac{1}{x_1^{(0)}} - \left[\frac{1}{x_1^{(0)}}\right] \right)$$

$$= (x_1^{(1)}, x_2^{(1)}, \ldots, x_n^{(1)}),$$

whereby $x_1 \neq 0$; for $x_1 = 0$ slight modifications have to be made (already observed by O. Perron and called "Stoerungen"). The successive transformations T^v are then characterized by the vectors

$$\left(\left[\frac{x_2^{(v)}}{x_1^{(v)}}\right], \left[\frac{x_3^{(v)}}{x_1^{(v)}}\right], \ldots, \left[\frac{x_n^{(v)}}{x_1^{(v)}}\right], \left[\frac{1}{x_1^{(v)}}\right] \right) = (b_1^{(v)}, b_2^{(v)}, \ldots, b_n^{(v)})$$

and the basic problem that Schweiger poses in connection with this process of transformations, whereby also $v \to \infty$ is possible, if not all components of $(x_1^{(0)}, \ldots, x_n^{(0)})$ are rational, is the following: what is the set $\mathfrak{M}_v$ of points $x^{(0)} \in E_n$ whose T^v successive transformations are characterized, up to the v-th step, by effectively given vectors $(b_1^{(i)}, b_2^{(i)}, \ldots, b_n^{(i)})$, $(i = 0, \ldots, v)$; if $\mathfrak{M}_v$ can thus be defined, what is the Lebesgue measure of $\mathfrak{M}_v$ or how can it best be approximated. Schweiger has given very good approximation formulas for this measure and, after having obtained this instrument, he attacked masterly all the other ergodic problems of this Jacobi "transformation". Ergodic theories were already previously applied, most successively, to number-theoretic questions by Yu V. Linnik [19]. Yet it remains doubtful whether algebraic questions as the periodicity problem of the Jacobi algorithm could be handled by the ergodic approach. For the set of algebraic vectors (viz. vectors with algebraic components) in E_n is countable; its Lebesgue measure is zero, and it is exactly the sets of measure zero which are treated as stepchildren by ergodic theory. The Jacobi algorithm is still waiting for the master-mind to decipher it completely.

Chapter 1.

BASIC CONCEPTS AND RELATIONS

§ 1. Definition of the JACOBI-PERRON Algorithm

Throughout this book, E_{n-1} will denote the n-1-dimensional Euclidean vector space of n-1-tuples as real numbers ($n \geqslant 2$). Since only denumerable sets of vectors in E_{n-1} will be investigated, we shall use the notation

$$(1.1) \qquad E_{n-1} \ni a^{(k)} = (a_1^{(k)}, a_2^{(k)}, \ldots, a_{n-1}^{(k)}) \qquad (a_i^{(k)} \in R; i=1,\ldots,n-1)$$

where k denotes a non-negative rational integer.

Definition I. By T the following transformation of E_{n-1} into E_{n-1} is meant: let

$$(1.2) \qquad f(a^{(k)}) = b^{(k)} = (b_1^{(k)}, \ldots, b_{n-1}^{(k)}) \in E_{n-1}$$

be any vector function on E_{n-1} with values in E_{n-1} such that

$$(1.3) \qquad a_1^{(k)} \neq b_1^{(k)},$$

then

$$(1.4) \qquad a^{(k)}T = (a_1^{(k)} - b_1^{(k)})^{-1} (a_2^{(k)} - b_2^{(k)}, \ldots, a_{n-1}^{(k)} - b_{n-1}^{(k)}, 1). \quad (k=0,1,\ldots)$$

We shall call $f(a^{(k)})$ a T-function (or a function associated with T).

Example 1. Let $a^{(0)} = (a_1^{(0)}, \ldots, a_{n-1}^{(0)})$, $a_i^{(0)} \neq 0$; $(i=1,\ldots,n-1)$ $f(a^{(k)}) = \frac{1}{2} a^{(k)}$ $(k=0,1,\ldots)$; then $a^{(0)}T = a_1^{-1}(a_2^{(0)}, \ldots, a_{n-1}^{(0)}, 2)$.

Definition II. A sequence $\langle a^{(k)} \rangle \equiv a^{(0)}, a^{(1)}, \ldots, a^{(k)}, \ldots$ of vectors in E_{n-1} will be called a Jacobi-Perron Algorithm (in short: JPA) of the vector $a^{(0)}$, if there exists a T-transformation of E_{n-1} into E_{n-1} such that for every k

$$(1.5) \qquad a^{(k)}T = a^{(k+1)}.$$

For successive T-transformations we use the customary notation

$$(1.6)\quad \begin{cases} (a^{(k)}T^{v})T = a^{(k)}T^{v+1}, & (v=1,2,\dots) \\ (a^{(k)}T^{v})T^{s} = a^{(k)}T^{v+s}, & (v,s=1,2,\dots) \\ a^{(k)}T^{v} = a^{(k+v)}. & (k=0,1,\dots;v=1,2,\dots) \end{cases}$$

Specifically, we have $a^{(0)}T^{k} = a^{(k)}$.

Example 2. $a^{(0)}$ and $f(a^{k})$ as in Example 1. One calculates easily

$$a^{(0)}T = \frac{1}{a_1^{(0)}}(a_2^{(0)},a_3^{(0)},\dots,a_{n-1}^{(0)},2);$$

$$a^{(0)}T^{2} = \frac{1}{a_2^{(0)}}(a_3^{(0)},a_4^{(0)},\dots,a_{n-1}^{(0)},2,2a_1^{(0)});$$

- -

$$a^{(0)}T^{n-2} = \frac{1}{a_{n-2}^{(0)}}(a_{n-1}^{(0)},2,2a_1^{(0)},\dots,2a_{n-3}^{(0)});$$

$$a^{(0)}T^{n-1} = \frac{1}{a_{n-1}^{(0)}}(2,2a_1^{(0)},\dots,2a_{n-2}^{(0)});$$

$$a^{(0)}T^{n} = (a_1^{(0)},a_2^{(0)},\dots,a_{n-1}^{(0)}).$$

We have thus obtained the formula

$$a^{(0)}T^{n} = a^{(0)},$$

and generally

$$a^{(0)}T^{sn} = a^{(0)},$$

$$a^{(0)}T^{sn+k} = \frac{1}{a_k^{(0)}}(a_{k+1}^{(0)},\dots,a_{n-1}^{(0)},2,2a_1^{(0)},\dots,2a_{k-1}^{(0)}),$$

$$a^{(0)}T^{sn+n-1} = \frac{1}{a_{n-1}^{(0)}}(2,2a_1^{(0)},\dots,2a_{n-2}^{(0)}).\quad (k-1,\dots,n-2;s=0,1,\dots)$$

§ 2. The Matrices of the JPA

By $A^{(v)}$ we shall denote the following successive matrices of the JPA: $A^{(0)}$ is the unit matrix

$$(1.7)\qquad A^{(0)} = (A_i^{(j)});\quad A_i^{(j)} = \delta_{ij}\quad (i,j=0,1,\ldots,n-1)$$

where δ_{ij} is the Kronecker delta.

We shall define numbers $A_i^{(j)}$ for higher indices j by

$$(1.8)\qquad A_i^{(n+v)} = \sum_{j=0}^{n-1} b_j^{(v)} A_i^{(v+j)}.\quad (b_0^{(v)} = a_0^{(v)} = 1;\ i=0,\ldots,n-1;\ v=0,1,\ldots)$$

We introduce the transformation matrices

$$(1.9)\qquad B^{(v)} = \begin{pmatrix} 0 & 0 & 0 & \cdots & 0 & 1 \\ 1 & 0 & . & \cdots & 0 & b_1^{(v)} \\ 0 & 1 & 0 & \cdots & 0 & b_2^{(v)} \\ \cdots & \cdots & \cdots & \cdots & \cdots & \cdots \\ 0 & \cdots & \cdots & 0 & 1 & b_{n-1}^{(v)} \end{pmatrix}.\quad (v=1,2,\ldots)$$

We then obtain the successive matrices $A^{(v)}$ from

$$(1.10)\qquad A^{(v+1)} = A^{(v)}B^{(v)} = (A_i^{(v+j)}).\quad (v=0,1,\ldots;\ i,j=0,\ldots,n-1)$$

Since

$$(1.11)\qquad \det B^{(v)} = (-1)^{n-1},\quad \det A^{(0)} = 1,$$

we obtain from (1.10), (1.11)

$$(1.12)\qquad \det A^{(v)} = (-1)^{v(n-1)}.\quad (v=0,1,\ldots)$$

We shall use formula (1.12) frequently.

Example 3. $a^{(0)} = (u,v,w);\ f(a^{(k)}) = \frac{1}{2}a^{(k)}.\quad (k=0,1,\ldots)$

We obtain successively from Example 2.

$$\begin{aligned} b^{(0)} &= \tfrac{1}{2}(u,v,w), \\ b^{(1)} &= \tfrac{1}{2u}(v,w,2), \\ b^{(2)} &= \tfrac{1}{2v}(w,2,2u), \\ b^{(3)} &= \tfrac{1}{2w}(2,2u,2v), \\ b^{(4)} &= \tfrac{1}{2}(u,v,w). \end{aligned}$$

$$
\begin{array}{llll}
A_0^{(0)} = 1 ; & A_1^{(0)} = 0 ; & A_2^{(0)} = 0 ; & A_3^{(0)} = 0 ; \\
A_0^{(1)} = 0 ; & A_1^{(1)} = 1 ; & A_2^{(1)} = 0 ; & A_3^{(1)} = 0 ; \\
A_0^{(2)} = 0 ; & A_1^{(2)} = 0 ; & A_2^{(2)} = 1 ; & A_3^{(2)} = 0 ; \\
A_0^{(3)} = 0 ; & A_1^{(3)} = 0 ; & A_2^{(3)} = 0 ; & A_3^{(3)} = 1 ; \\
A_0^{(4)} = 1 ; & A_1^{(4)} = \frac{u}{2} ; & A_2^{(4)} = \frac{v}{2} ; & A_3^{(4)} = \frac{w}{2} ; \\
A_0^{(5)} = \frac{1}{u} ; & A_1^{(5)} = \frac{3}{2} ; & A_2^{(5)} = \frac{v}{u} ; & A_3^{(5)} = \frac{w}{4} ; \\
A_0^{(6)} = \frac{2}{v} ; & A_1^{(6)} = \frac{2u}{v} ; & A_2^{(6)} = \frac{5}{2} ; & A_3^{(6)} = \frac{2w}{v} ; \\
A_0^{(7)} = \frac{4}{w} ; & A_1^{(7)} = \frac{4u}{w} ; & A_2^{(7)} = \frac{4v}{w} ; & A_3^{(7)} = \frac{9}{2} ; \\
A_0^{(8)} = \frac{9}{2} ; & A_1^{(8)} = \frac{17u}{4} ; & A_2^{(8)} = \frac{17v}{4} ; & A_3^{(8)} = \frac{17w}{4} ; \\
A_0^{(9)} = \frac{17}{2u} . & A_1^{(9)} = \frac{35}{4} ; & A_2^{(9)} = \frac{17v}{2u} ; & A_3^{(9)} = \frac{17w}{2u} .
\end{array}
$$

We verify easily

$$
\det A^{(2)} = \begin{vmatrix} 0 & 0 & 1 & \frac{1}{u} \\ 0 & 0 & \frac{u}{2} & \frac{3}{2} \\ 1 & 0 & \frac{v}{2} & \frac{v}{u} \\ 0 & 1 & \frac{w}{2} & \frac{w}{u} \end{vmatrix} = 1 = (-1)^{2\cdot 3}
$$

$$
\det A^{(3)} = \begin{vmatrix} 0 & 1 & \frac{1}{u} & \frac{2}{v} \\ 0 & \frac{u}{2} & \frac{3}{2} & \frac{2u}{v} \\ 0 & \frac{v}{2} & \frac{v}{u} & \frac{5}{2} \\ 1 & \frac{w}{2} & \frac{w}{u} & \frac{2w}{v} \end{vmatrix} = -1 = (-1)^{3\cdot 3}
$$

$$\det A^{(4)} = \begin{vmatrix} 1 & \frac{1}{u} & \frac{2}{v} & \frac{4}{w} \\ \frac{u}{2} & \frac{3}{2} & \frac{2u}{v} & \frac{4u}{w} \\ \frac{v}{2} & \frac{v}{u} & \frac{5}{2} & \frac{4v}{w} \\ \frac{w}{2} & \frac{w}{u} & \frac{2w}{v} & \frac{9}{2} \end{vmatrix} = 1 = (-1)^{4\cdot 3}$$

$$\det A^{(5)} = \begin{vmatrix} \frac{1}{u} & \frac{2}{v} & \frac{4}{w} & \frac{9}{2} \\ \frac{3}{2} & \frac{2u}{v} & \frac{4u}{w} & \frac{17u}{4} \\ \frac{v}{u} & \frac{5}{2} & \frac{4v}{w} & \frac{17v}{4} \\ \frac{w}{u} & \frac{2w}{v} & \frac{9}{2} & \frac{17w}{4} \end{vmatrix} = -1 = (-1)^{5\cdot 3}$$

§ 3. Basic Relations

By means of the entries of the matrices $A^{(v)}$ important relations can now be derived. By definition of (1.4), we express every vector $a^{(k)}$ as a function of its successor, viz.

$$(1.13) \qquad a_{i+1}^{(k)} = b_{i+1}^{(k)} + \frac{a_i^{(k+1)}}{a_{n-1}^{(k+1)}}. \qquad (i=0,1,\ldots,n-2)$$

We remind the reader that $a_0^{(k)} = 1$ $(k=0,1,\ldots)$ by definition. We now prove the basic formula

$$(1.14) \qquad a^{(0)} = \left(\sum_{j=0}^{n-1} a_j^{(v)} A_0^{(v+j)}\right)^{-1} \times \left(\sum_{j=0}^{n-1} a_j^{(v)} A_1^{(v+j)}, \sum_{j=0}^{n-1} a_j^{(v)} A_2^{(v+j)}, \ldots, \sum_{j=0}^{n-1} a_j^{(v)} A_{n-1}^{(v+j)}\right).$$

$$(v=0,1,2,\ldots)$$

Proof by induction. (1.14) is correct for $v = 0$, as can be easily verified; we obtain, on basis of (1.13),

$$a^{(0)} =$$

$$\left(A_0^{(v)}+\sum_{j=0}^{n-2} a_{j+1}^{(v)} A_0^{(v+1+j)}\right)^{-1}\left(A_1^{(v)}+\sum_{j=0}^{n-2} a_{j+1}^{(v)} A_1^{(v+1+j)},\dots, A_{n-1}^{(v)}+\sum_{j=0}^{n-2} a_{j+1}^{(v)} A_{n-1}^{(v+1+j)}\right)$$

$$= \left(A_0^{(v)} +\sum_{j=0}^{n-2} \left(b_{j+1}^{(v)} + \frac{a_j^{(v+1)}}{a_{n-1}^{(v+1)}}\right) A_0^{(v+1+j)}\right)^{-1}$$

$$\times \left(A_1^{(v)} + \sum_{j=0}^{n-2} \left(b_{j+1}^{(v)} + \frac{a_j^{(v+1)}}{a_{n-1}^{(v+1)}}\right) A_1^{(v+1+j)}, \dots, A_{n-1}^{(v)}+\sum_{j=0}^{n-2}\left(b_{j+1}^{(v)} + \frac{a_j^{(v+1)}}{a_{n-1}^{(v+1)}}\right) A_{n-1}^{(v+1+j)}\right)$$

$$= \left(\sum_{j=0}^{n-1} b_j^{(v)} A_0^{(v+j)} + \sum_{j=0}^{n-2} \frac{a_j^{(v+1)} A_0^{(v+1+j)}}{a_{n-1}^{(v+1)}}\right)^{-1} \times$$

$$\left(\sum_{j=0}^{n-1} b_j^{(v)} A_1^{(v+j)}+\sum_{j=0}^{n-1} \frac{a_j^{(v+1)} A_1^{(v+1+j)}}{a_{n-1}^{(v+1)}},\dots,\sum_{j=0}^{n-1} b_j^{(v)} A_{n-1}^{(v+j)}+\sum_{j=0}^{n-2} \frac{a_j^{(v+1)} A_{n-1}^{(v+1+j)}}{a_{n-1}^{(v+1)}}\right)$$

$$=\left(\sum_{j=0}^{n-1} a_j^{(v+1)} A_0^{(v+1+j)}\right)^{-1}\left(\sum_{j=0}^{n-1} a_j^{(v+1)} A_1^{(v+1+j)},\dots,\sum_{j=0}^{n-1} a_j^{(v+1)} A_{n-1}^{(v+1+j)}\right) .$$

For further purposes we shall need the formula

$$\text{(1.15)} \qquad \begin{vmatrix} 1 & A_0^{(v+1)} & \dots & A_0^{(v+n-1)} \\ a_1^{(0)} & A_1^{(v+1)} & \dots & A_1^{(v+n-1)} \\ \vdots & \vdots & & \vdots \\ a_{n-1}^{(0)} & A_{n-1}^{(v+1)} & \dots & A_{n-1}^{(v+n-1)} \end{vmatrix} = \left(\sum_{j=0}^{n-1} a_j^{(v)} A_0^{(v+j)}\right)^{-1} (-1)^{v(n-1)}.$$

$$(v=0,1,\dots)$$

Proof of (1.15). Substituting for the components of $a^{(0)}$ their values from (1.14) in the first column of the $n \times n$ matrix in (1.15), we obtain

$$\begin{vmatrix} 1 & A_0^{(v+1)} & \cdots & A_0^{(v+n-1)} \\ a_1^{(0)} & A_1^{(v+1)} & \cdots & A_1^{(v+n-1)} \\ \vdots & \vdots & & \vdots \\ a_{n-1}^{(0)} & A_{n-1}^{(v+1)} & \cdots & A_{n-1}^{(v+n-1)} \end{vmatrix} =$$

$$\begin{vmatrix} 1, & A_0^{(v+1)}, & \ldots, & A_0^{(v+n-1)} \\ \left(\sum_{j=0}^{n-1} a_j^{(v)} A_0^{(v+j)}\right)^{-1}\left(\sum_{j=0}^{n-1} a_j^{(v)} A_1^{(v+j)}\right), & A_1^{(v+1)}, & \ldots, & A_1^{(v+n-1)} \\ \vdots & \vdots & & \vdots \\ \left(\sum_{j=0}^{n-1} a_j^{(v)} A_0^{(v+j)}\right)^{-1}\left(\sum_{j=0}^{n-1} a_j^{(v)} A_{n-1}^{(v+j)}\right), & A_{n-1}^{(v+1)}, & \ldots, & A_{n-1}^{(v+n-1)} \end{vmatrix}$$

$$= \left(\sum_{j=0}^{n-1} a_j^{(v)} A_0^{(v+j)}\right)^{-1} \begin{vmatrix} \sum_{j=0}^{n-1} a_j^{(v)} A_0^{(v+j)}, & A_0^{(v+1)}, & \ldots, & A_0^{(v+n-1)} \\ \sum_{j=0}^{n-1} a_j^{(v)} A_1^{(v+j)}, & A_1^{(v+1)}, & \ldots, & A_1^{(v+n-1)} \\ \vdots & \vdots & & \vdots \\ \sum_{j=0}^{n-1} a_j^{(v)} A_{n-1}^{(v+j)}, & A_{n-1}^{(v+1)}, & \ldots, & A_{n-1}^{(v+n-1)} \end{vmatrix}$$

$$= \left(\sum_{j=0}^{n-1} a_j^{(v)} A_0^{(v+j)}\right)^{-1} \begin{vmatrix} A_0^{(v)} & A_0^{(v+1)} & \cdots & A_0^{(v+n-1)} \\ A_1^{(v)} & A_1^{(v+1)} & \cdots & A_1^{(v+n-1)} \\ \vdots & \vdots & & \vdots \\ A_{n-1}^{(v)} & A_{n-1}^{(v+1)} & \cdots & A_{n-1}^{(v+n-1)} \end{vmatrix}$$

$$= \left(\sum_{j=0}^{n-1} a_j^{(v)} A_0^{(v+j)}\right)^{-1} (-1)^{v(n-1)} .$$

We shall conclude this paragraph with proving formula

$$\prod_{i=1}^{v} a_{n-1}^{(i)} = \sum_{j=0}^{n-1} a_j^{(v)} A_0^{(v+j)} . \quad (v=1,2,\ldots) \tag{1.16}$$

Proof by induction. (1.16) is correct for $v=1$, since then $A_0^{(1)} = A_0^{(2)} = \cdots = A_0^{(n-1)} = 0$; $A_0^{(n)} = 1$. We further obtain, on basis of (1.13),

$$\prod_{i=1}^{v} a_{n-1}^{(i)} = A_0^{(v)} + \sum_{j=0}^{n-2} \left(b_{j+1}^{(v)} + \frac{a_j^{(v+1)}}{a_{n-1}^{(v+1)}} \right) A_0^{(v+1+j)}$$

$$= \left(\sum_{j=0}^{n-1} b_j^{(v)} A_0^{(v+j)} \right) + \sum_{j=0}^{n-2} \frac{a_j^{(v+1)}}{a_{n-1}^{(v+1)}} A_0^{(v+1+j)}$$

$$= A_0^{(v+n)} + \sum_{n=0}^{n-2} \frac{a_j^{(v+1)}}{a_{n-1}^{(v+1)}} A_0^{(v+1+j)}$$

$$= \frac{\sum_{j=0}^{n-1} a_j^{(v+1)} A_0^{(v+1+j)}}{a_{n-1}^{(v+1)}} ;$$

so $a_{n-1}^{(v+1)} \prod_{i=1}^{v} a_{n-1}^{(i)} = \sum_{j=0}^{n-1} a_j^{(v+1)} A_0^{(v+1+j)}$.

Chapter 2.

CONVERGENCE OF JPA

§ 1. An Analogy with Continued Fractions

To introduce the concept of convergence of the JPA, we shall recall convergence of continued fractions. For $n = 2$, E_{n-1} becomes the real number space; the JPA becomes the algorithm of continued fractions, yet we shall not alter the notations of Chapter 1. The formulas to be used here take the more simple form

$$a_1^{(k)} = b_1^{(k)} + \frac{1}{a_1^{(k+1)}}; \qquad (k=0,1,\ldots) \tag{2.1}$$

$$b_1^{(k)} = f(a_1^{(k)}); \qquad (k=0,1,\ldots) \tag{2.2}$$

$$\begin{gathered} A_0^{(0)} = 1, \quad A_0^{(1)} = 0; \quad A_1^{(0)} = 0; \quad A_1^{(1)} = 1; \\ A_i^{(v+2)} = A_i^{(v)} + b_1^{(v)} A_i^{(v+1)}. \quad (i=0,1;\ v=0,1,\ldots) \end{gathered} \tag{2.3}$$

$$\begin{vmatrix} A_0^{(v)} & A_0^{(v+1)} \\ A_1^{(v)} & A_1^{(v+1)} \end{vmatrix} = (-1)^v; \qquad (v=0,1,\ldots) \tag{2.4}$$

$$a_1^{(0)} = \frac{A_1^{(v)} + b_1^{(v)} A_1^{(v+1)}}{A_0^{(v)} + b_0^{(v)} A_0^{(v+1)}}. \tag{2.5}$$

We shall now presume boundedness of $f(a^k)$ such that

$$1 \leqslant f(a_1{}^k) < a_1{}^{(k)} < \infty. \qquad (k=1,2,\ldots) \tag{2.6}$$

The nature of this restriction on $f(a_1{}^{(k)})$ will be explained later. We obtain from (2.6), on basis of (2.3)

$$\lim_{v \to \infty} A_0^{(v)} = +\infty; \quad A_0^{(v)} < +\infty. \qquad (v=0,1,\ldots) \tag{2.7}$$

We can now prove easily the well-known formula for continued fractions of approximating a real number by its convergents

(2.8) $$a_1^{(0)} = \lim_{v\to\infty} \frac{A_1^{(v)}}{A_0^{(v)}} .$$

From (2.3),...,(2.7) we obtain, for $v > 1$,

$$\left| a_1^{(0)} - \frac{A_1^{(v)}}{A_0^{(v)}} \right| = \left| \frac{A_1^{(v-1)} + a_1^{(v-1)} A_1^{(v)}}{A_0^{(v-1)} + a_1^{(v-1)} A_0^{(v)}} - \frac{A_1^{(v)}}{A_0^{(v)}} \right|$$

$$= \frac{|A_0^{(v-1)} A_1^{(v)} - A_0^{(v)} A_1^{(v-1)}|}{A_0^{(v)} (A_0^{(v-1)} + a_1^{(v-1)} A_0^{(v)})} = \frac{1}{A_0^{(v)} (A_0^{(v-1)} + a_1^{(v-1)} A_0^{(v)})} < \frac{1}{A_0^{(v)2}} ;$$

$$\lim_{v\to\infty} \left| a_1^{(0)} - \frac{A_1^{(v)}}{A_0^{(v)}} \right| = 0.$$

Example 4. $a^{(0)} = a;\ f(a^{(k)}) = \frac{1}{2} a^{(k)}.$ $(a > 0;\ k = 0,1,\ldots)$

We obtain from Example 3.

$$f(a^{(2k)}) = \frac{a}{2};\quad f(a^{(2k+1)}) = \frac{1}{a},$$

and from (1.8)

$$A_i^{(2n+2)} - A_i^{(2n)} = \frac{a}{2} A_i^{(2n+1)}. \qquad (i=0,1)$$

The reader should note that for no real number $a > 0$ is the condition $f(a_1^{(k)}) \geqslant 1$ fulfilled for every k, though the condition $f(a_1^{(k)}) < a^{(k)}$ holds for every k; the restriction $1 \leqslant f(a_1^{(k)}) < a_1^{(k)} < +\infty$ is, therefore, only a sufficient condition for the convergence of the algorithm in the case $n = 2$; we used it for showing that $A_0^{(v)-2} \to 0$. This yields

$$A_i^{(2n+2)} - A_i^{(0)} = \frac{a}{2} \left[A_i^{(2n+1)} + A_i^{(2n-1)} + \cdots + A_i^{(1)} \right],$$

$$\frac{1}{a} A_i^{(2n+2)} - \frac{1}{a} A_i^{(0)} = \frac{1}{2} A_i^{(2n+1)} + \frac{1}{2} (A_i^{(2n-1)} + A_i^{(2n-3)} + \cdots + A_i^{(1)}),$$

$$A_i^{(2n+1)} + \frac{1}{a} A_i^{(2n+2)} - \frac{1}{a} A_i^{(0)} = \frac{3}{2} A_i^{(2n+1)} + \frac{1}{2} (A_i^{(2n-1)} + A_i^{(2n-3)} + \cdots + A_i^{(1)}),$$

$$A_i^{(2n+3)} - \frac{1}{a} A_i^{(0)} = \frac{3}{2} A_i^{(2n+1)} + \frac{1}{2} (A_i^{(2n-1)} + A_i^{(2n-3)} + \cdots + A_i^{(1)}),$$

$$A_i^{(2n+1)} - \frac{1}{a} A_i^{(0)} = \frac{3}{2} A_i^{(2n-1)} + \frac{1}{2}(A_i^{(2n-3)} + A_i^{(2n-5)} + \cdots + A_i^{(1)});$$

the latter equation was obtained from the preceding one by substituting n-1 for n; subtracting:

$$A_i^{(2n+3)} - A_i^{(2n+1)} = \frac{3}{2} A_i^{(2n+1)} - A_i^{(2n-1)},$$

$$A_i^{(2n+3)} - 2\,A_i^{(2n+1)} = \frac{1}{2}(A_i^{(2n+1)} - 2\,A_i^{(2n-1)}).$$

From this equation we obtain, by known techniques:

$$A_i^{(2n+3)} - 2\,A_i^{(2n+1)} = \frac{1}{2^n}(A_i^{(3)} - 2\,A_i^{(1)}).$$

For i=0,1, we obtain explicit formulas (after easy operations) for $A_0^{(2n+1)}$, $A_1^{(2n+1)}$. Taking as our initial equation the relation

$$A_i^{(2n+3)} - A_i^{(2n+1)} = \frac{1}{a} A_i^{(2n)},$$

we obtain directly explicit formulas for $A_0^{(2n)}$, $A_1^{(2n)}$; the results are

$$A_0^{(2n)} = \frac{2^n + 2^{-n+1}}{3} = \frac{1}{3}(2^n + 2^{-(n-1)})$$

$$A_0^{(2n+1)} = \frac{2^{2n} - 1}{3a\cdot 2^{n-1}} = \frac{1}{3a}(2^{n+1} - 2^{-(n-1)})$$

$$A_1^{(2n)} = \frac{a}{3}(2^n - 2^{-n});$$

$$A_1^{(2n+1)} = \frac{2^{2n+1} + 1}{3\cdot 2^n} = \frac{1}{3}(2^{n+1} + 2^{-n}).$$

We now verify easily

$$\lim_{n\to\infty} \frac{A_1^{(2n)}}{A_0^{(2n)}} = \lim_{n\to\infty} \frac{A_1^{(2n+1)}}{A_0^{(2n+1)}} = a.$$

§ 2. The First Main Convergence Criterion of JPA

In analogy to the convergence of continued fractions we now formulate

<u>Definition III.</u> Let $a^{(0)} \in E_{n-1}$; the JPA of $a^{(0)}$ is said to be convergent, if

$$a_i = \lim_{v\to\infty} \frac{A_i^{(v)}}{A_0^{(v)}} . \qquad (i=1,\dots,n-1)$$

<u>Definition IV.</u> A T-function

$$f(a^k) = b^{(k)} = (b_1^{(k)}, b_2^{(k)}, \dots, b_{n-1}^{(k)}),$$

is said to be P-bounded, if

$$0 < \frac{1}{b_{n-1}^{(k)}} \leqslant C, \; 0 \leqslant \frac{b_i^{(k)}}{b_{n-1}^{(k)}} \leqslant C, \quad (i=1,\dots,n-2;\; k=0,1,2,\dots)$$

where C is a real constant, independent of k.

These conditions were first introduced by O. Perron [22].

Example 5. Let again $a^{(0)} = (u,v) \in E_2$; $u,v > 0$; $f(a^{(k)}) = \frac{1}{2}a^{(k)}$. We obtain by direct calculation, or from Example 2,

$$b^{(3k)} = (\frac{u}{2}, \frac{v}{2}),$$

$$b^{(3k+1)} = (\frac{v}{2u}, \frac{1}{u}),$$

$$b^{(3k+2)} = (\frac{1}{v}, \frac{u}{v}), \qquad (k=0,1,\dots)$$

and it is now easily verified that max $[u, \frac{1}{u}, \frac{v}{2}, \frac{2}{v}, \frac{u}{v}, \frac{v}{u}]$ supplies such a C as demanded in Definition IV. We shall now state the First Main Convergence Criterion of JPA in

THEOREM 1. Let $a^{(0)} \in E_{n-1}$; the JPA of $a^{(0)}$ is convergent, if its T-function $f(a^k)$ is P-bounded.

Proof. For $v \geqslant n$, the numbers $A_0^{(v)}$ are all positive. This is a result of the P-boundedness of $f(a^k)$, viz.

$$A_0^{(n)} = 1, \; A_0^{(n+1)} = b_{n-1}^{(1)}; \quad A_0^{(n+2)} = b_{n-2}^{(2)}A_0^{(n)} + b_{n-1}^{(2)}A_0^{(n+1)},$$

$$A_0^{(n+2)} \geqslant b_{n-1}^{(1)}b_{n-1}^{(2)}, \; A_0^{(n+3)} = b_{n-3}^{(3)}A_0^{(n)} + b_{n-2}^{(3)}A_0^{(n+1)} + b_{n-1}^{(3)}A_0^{(n+2)},$$

$$A_0^{(n+3)} \geqslant b_{n-1}^{(1)} b_{n-1}^{(2)} b_{n-1}^{(3)},$$

and, by induction,

$$A_0^{(n+v)} \geqslant \prod_{i=0}^{v} b_{n-1}^{(i)} > 0. \qquad (v=0,1,\ldots) \tag{2.9}$$

Let denote

$$m_i^{(v)} = \min\left[\frac{A_i^{(v)}}{A_0^{(v)}}, \frac{A_i^{(v+1)}}{A_0^{(v+1)}}, \ldots, \frac{A_i^{(v+n-1)}}{A_0^{(v+n-1)}}\right],$$

(2.10)

$$M_i^{(v)} = \max\left[\frac{A_i^{(v)}}{A_0^{(v)}}, \frac{A_i^{(v+1)}}{A_0^{(v+1)}}, \ldots, \frac{A_i^{(v+n-1)}}{A_0^{(v+n-1)}}\right]. \qquad (v=n,n+1,\ldots)$$

We further obtain, on basis of (1.8),

$$\frac{A_i^{(v+n)}}{A_0^{(v+n)}} = \frac{A_i^{(v)} + b_1^{(v)} A_i^{(v+1)} + b_2^{(v)} A_i^{(v+2)} + \cdots + b_{n-1}^{(v)} A_i^{(v+n-1)}}{A_0^{(v+n)}} =$$

$$= \frac{A_i^{(v)}}{A_0^{(v)}} \frac{A_0^{(v)}}{A_0^{(v+n)}} + \frac{A_i^{(v+1)}}{A_0^{(v+1)}} \cdot \frac{b_1^{(v)} A_0^{(v+1)}}{A_0^{(v+n)}} + \cdots + \frac{A_i^{(v+n-1)}}{A_0^{(v+n-1)}} \frac{b_{n-1}^{(v)} A_0^{(v+n-1)}}{A_0^{(v+n)}} .$$

Putting

$$t_i^{(v)} = \frac{b_i^{(v)} A_0^{(v+i)}}{A_0^{(v+n)}}, \qquad (i=0,\ldots,n-1) \tag{2.11}$$

we obtain

$$\frac{A_i^{(v+n)}}{A_0^{(v+n)}} = t_0^{(v)} \frac{A_i^{(v)}}{A_0^{(v)}} + t_1^{(v)} \frac{A_i^{(v+1)}}{A_0^{(v+1)}} + \cdots + t_{n-1}^{(v)} \frac{A_i^{(v+n-1)}}{A_0^{(v+n-1)}} . \tag{2.12}$$

But

$$A_0^{(v+n)} = A_0^{(v)} + b_1^{(v)} A_0^{(v+1)} + \cdots + b_i^{(v)} A_0^{(v+1)} + b_{n-1}^{(v)} A_0^{(v+n-1)},$$

so that $0 \leqslant t_i^{(v)} \leqslant 1$, and from (2.12), in virtue of (2.10),

$$\frac{A_i^{(v+n)}}{A_0^{(v+n)}} \geqslant t_0^{(v)} m_i^{(v)} + t_1^{(v)} m_i^{(v)} + \cdots + t_{n-1}^{(v)} m_i^{(v)},$$

$$\frac{A_i^{(v+n)}}{A_0^{(v+n)}} \leqslant t_0^{(v)} M_i^{(v)} + t_1^{(v)} M_i^{(v)} + \cdots + t_{n-1}^{(v)} M_i^{(v)}.$$

Since

$$t_0^{(v)} + t_1^{(v)} + \cdots + t_{n-1}^{(v)} = \frac{A_0^{(v)}}{A_0^{(v+n)}} + \frac{b_1^{(v)} A_0^{(v+1)}}{A_0^{(v+n)}} + \cdots + \frac{b_{n-1}^{(v)} A_0^{(v+n-1)}}{A_0^{(v+n)}}$$

$$= \frac{A_0^{(v+n)}}{A_0^{(v+n)}} = 1,$$

we finally obtain

(2.13)
$$\frac{A_i^{(v+n)}}{A_0^{(v+n)}} \geqslant m_i^{(v)}, \quad \frac{A_i^{(v+n)}}{A_0^{(v+n)}} \leqslant M_i^{(v)}.$$

From

$$m_i^{(v+1)} = \min\left(\frac{A_i^{(v+1)}}{A_0^{(v+1)}}, \ldots, \frac{A_i^{(v+n-1)}}{A_0^{(v+n-1)}}, \frac{A_i^{(v+n)}}{A_0^{(v+n)}}\right)$$

$$= \min\left(m_i^{(v)}, \frac{A_i^{(v+n)}}{A_0^{(v+n)}}\right),$$

we obtain, in virtue of (2.13)

(2.14)
$$m_i^{(v+1)} \geqslant m_i^{(v)},$$

and similarly

(2.15)
$$M_i^{(v+1)} \leqslant M_i^{(v)}.$$

We thus have the chain of inequalities

(2.16)
$$0 \leqslant m_i^{(v)} \leqslant m_i^{(v+1)} \leqslant M_i^{(v+1)} < M_i^{(v)}.$$

Thus $\langle m_i^{(v)} \rangle$ is a non-decreasing sequence bounded from above, and

$\langle M_i^{(v)} \rangle$ is a non-increasing sequence bounded from below (for $v > n$) and their limits exist; denoting these by

$$(2.17) \qquad \lim_{v\to\infty} m_i^{(v)} = m_i < \infty, \qquad \lim_{v\to\infty} M_i^{(v)} = M_i < \infty$$

we have

$$(2.18) \qquad 0 \leqslant m_i^{(v)} \leqslant m_i \leqslant M_i \leqslant M_i^{(v)}. \qquad (v=n,n+1,\dots)$$

There exists a rational number v_0, such that for every $\varepsilon > 0$ and $v > v_0$

$$m_i^{(v)} > m_i - \varepsilon \quad \text{which implies}$$

$$(2.19) \qquad \frac{A_i^{(v+j)}}{A_0^{(v+j)}} > m_i - \varepsilon . \qquad (j=0,1,\dots,n-1)$$

From (2.12), (2.19) we now obtain

$$\frac{A_i^{(v+n)}}{A_0^{(v+n)}} > (t_0^{(v)}+t_1^{(v)}+\dots+t_{n-2}^{(v)})(m_i-\varepsilon) + t_{n-1}^{(v)} \frac{A_i^{(v+n-1)}}{A_0^{(v+n-1)}}$$

$$= (1-t_{n-1}^{(v)})(m_i-\varepsilon) + t_{n-1}^{(v)} \frac{A_i^{(v+n-1)}}{A_0^{(v+n-1)}}$$

$$= m_i - (1-t_{n-1}^{(v)})\varepsilon + t_{n-1}^{(v)} \left(\frac{A_i^{(v+n-1)}}{A_0^{(v+n-1)}} - m_i \right)$$

$$\geqslant m_i - \varepsilon + t_{n-1}^{(v)} \left(\frac{A_i^{(v+n-1)}}{A_0^{(v+n-1)}} - m_i \right),$$

$$(2.20) \qquad \frac{A_i^{(v+n)}}{A_0^{(v+n)}} - m_i > t_{n-1}^{(v)} \left(\frac{A_i^{(v+n-1)}}{A_0^{(v+n-1)}} - m_i \right) - \varepsilon .$$

By condition of the theorem

$$A_0^{(v+n)} \geqslant b_{n-1}^{(v)} A_0^{(v+n-1)} \geqslant \frac{1}{C} A_0^{(v+n-1)}$$

$$\geqslant \frac{1}{C^2} A_0^{(v+n-2)} \geqslant \cdots \geqslant \frac{1}{C^s} A_0^{(v+n-s)};$$

from (2.11) we obtain, for $i=0,1,\ldots,n-2$;

$$\frac{t_i^{(v)}}{t_{n-1}^{(v)}} = \frac{b_i^{(v)} A_0^{(v+i)}}{b_{n-1}^{(v)} A_0^{(v+n-1)}} \leqslant C \cdot C^{n-i-1},$$

(2.21) $$\frac{t_i^{(v)}}{t_{n-1}^{(v)}} \leqslant C^{n-i}. \qquad (i=0,1,\ldots,n-2)$$

Thus

$$1 = t_0^{(v)} + t_1^{(v)} + \cdots + t_{n-2}^{(v)} + t_{n-1}^{(v)}$$

$$\leqslant (C^n + C^{n-1} + \cdots + C^2 + 1) t_{n-1}^{(v)},$$

(2.22) $$t_{n-1}^{(v)} > \frac{1}{1 + C + \cdots + C^n} = K.$$

From (2.20), (2.22) we now obtain

(2.23) $$\frac{A_i^{(v+n)}}{A_0^{(v+n)}} - m_i > K \left(\frac{A_i^{(v+n-1)}}{A_0^{(v+n-1)}} - m_i \right) - \varepsilon .$$

Since (2.23) holds for any $v > v_0$, we also have

$$\frac{A_i^{(v+n+1)}}{A_0^{(v+n+1)}} - m_i > K \left(\frac{A_i^{(v+n)}}{A_0^{(v+n)}} - m_i \right) - \varepsilon$$

$$> K \left[K \left(\frac{A_i^{(v+n-1)}}{A_0^{(v+n-1)}} - m_i \right) - \varepsilon \right] - \varepsilon$$

$$= K^2 \left(\frac{A_i^{(v+n-1)}}{A_0^{(v+n-1)}} - m_i \right) - \varepsilon - K\varepsilon,$$

and generally

$$\text{(2.24)} \qquad \frac{A_i^{(v+n+h)}}{A_0^{(v+n+h)}} - m_i > K^{h+1}\left(\frac{A_i^{(v+n-1)}}{A_0^{(v+n-1)}} - m_i\right) - \varepsilon(1 + K + \cdots + K^h).$$

Let $M_i^{(v')} = \max\left[\frac{A_i^{(v')}}{A_0^{(v')}}, \frac{A_i^{(v'+1)}}{A_0^{(v')}}, \ldots, \frac{A_i^{(v'+n-1)}}{A_0^{(v'+n-1)}}\right].$

Among the n numbers in the brackets, one must equal $M_i^{(v')}$, say

$$M_i^{(v')} = \frac{A_i^{(v'+j)}}{A_0^{(v'+j)}} .$$

We now specify v so that

$$v+n-1 = v'+j,$$

to obtain

$$\frac{A_i^{(v+n-1)}}{A_0^{(v+n-1)}} = M_i^{(v')} .$$

We now specify h so that

$$\frac{A_i^{(v+n+h)}}{A_0^{(v+n+h)}} = m_i^{(v'')} .$$

Recalling that $m_i = \sup\left\{m_i^{(v'')} \mid v'' > v_0\right\}$, $M_i = \inf\left\{M_i^{(v')} \mid v' > v_0\right\}$,

we now obtain from (2.24)

$$m_i - m_i > m_i^{(v'')} - m_i = \frac{A_i^{(v+n+h)}}{A_0^{(v+n+h)}} - m_i$$

$$> K^{h+1}\left(M_i^{(v')} - m_i\right) - \varepsilon(1+K+\cdots+K^h)$$

$$\geqslant K^{h+1}(M_i - m_i) - \varepsilon(1+K+\cdots+K^h);$$

$$0 > K^{(h+1)}(M_i - m_i) - \varepsilon(1+K+\cdots+K^h),$$

$$\text{(2.25)} \qquad \varepsilon(1+K+\cdots+K^h) > K^{h+1}(M_i - m_i).$$

By (2.18) $m_i \leqslant M_i$; if $M_i > m_i$, then (2.25) cannot hold, since ε may become arbitrarily small, while K, h, m_i, M_i are constant. Thus

$$(2.26) \qquad m_i = M_i = \lim_{v\to\infty} \frac{A_i^{(v)}}{A_0^{(v)}} .$$

Theorem 1 is now easily proved. We obtain, in virtue of (1.14)

$$(2.27) \qquad a_i^{(0)} = \frac{A_0^{(v)}}{\sum_{j=0}^{n-1} a_j^{(v)} A_0^{(v+j)}} \cdot \frac{A_i^{(v)}}{A_0^{(v)}} + \frac{A_0^{(v+1)} a_1^{(v)}}{\sum_{j=0}^{n-1} a_j^{(v)} A_0^{(v+j)}} \cdot \frac{A_i^{(v+1)}}{A_0^{(v+1)}} + \cdots$$

$$+ \frac{A_0^{(v+n-1)} a_{n-1}^{(v)}}{\sum_{j=0}^{n-1} a_j^{(v)} A_0^{(v+j)}} \cdot \frac{A_i^{(v+n-1)}}{A_0^{(v+n-1)}} ,$$

and, denoting

$$(2.28) \qquad \frac{A_0^{(v+j)} a_j^{(v)}}{\sum_{j=0}^{n-1} a_j^{(v)} A_0^{(v+j)}} = f_j^{(v)} \qquad (j=0,\ldots,n-1),$$

we obtain, as can be easily verified,

$$(2.29) \qquad 0 < f_j^{(v)} \leqslant 1, \quad f_0^{(v)} + f_1^{(v)} + \cdots + f_{n-1}^{(v)} = 1.$$

(2.27), (2.29) yield

$$(2.30) \qquad a_i^{(0)} = f_0^{(v)} \frac{A_i^{(v)}}{A_0^{(v)}} + f_1^{(v)} \frac{A_i^{(v+1)}}{A_0^{(v+1)}} + \cdots + f_{n-1}^{(v)} \frac{A_i^{(v+n-1)}}{A_0^{(v+n-1)}} .$$

Because of (2.26) we can put

$$(2.31) \qquad \frac{A_i^{(v+j)}}{A_0^{(v+j)}} = m_i - \varepsilon_j, \qquad (j=0,\ldots,n-1)$$

and obtain from (2.30), (2.31)

$$a_i^{(0)} = \sum_{j=0}^{n-1} f_j^{(v)} (m_i - \varepsilon_j) = m_i \sum_{j=0}^{n-1} f_j^{(v)} - \sum f_j^{(v)} \varepsilon_j$$

$$= m_i - \sum f_j^{(v)} \varepsilon_j ,$$

and from this equation, since the ε_j can be chosen arbitrarily small and $a_i^{(0)}$, m_i are constant, while the $f_j^{(v)}$ are bounded,

$$a_i^{(0)} = m_i = \lim_{v \to \infty} \frac{A_i^{(v)}}{A_0^{(v)}} .$$

Chapter 3.

PERIODICITY OF JPA

§1. Two Definitions of Periodicity

In this chapter a more rigorous definition of periodicity of JPA will be given, than is usually practiced by various authors. A strict observance of the following situation is often neglected: the JPA of a point $a^{(0)} \in E_{n-1}$ is completely characterized by its T-function $f(a^k)$. This function being well defined, the JPA may ultimately become periodic in the sense that a finite sequence of successive vectors $a^{(k)}$ will recur periodically, as was demonstrated in Example 2. Of course, if the JPA of $a^{(0)} \in E_{n-1}$ is carried out with a different T-function, its successive vectors $a^{(k)}$ will be entirely different, and the JPA may not become periodic, or may become so in a different pattern. Another point which often escapes attention is the following: periodicity is sometimes defined by recurrency of the vectors $a^{(k)}$, and in other places by that of the vectors $b^{(k)}$. That these two definitions are equivalent, is proved in this chapter.

Definition V. Let $a^{(0)} \in E_{n-1}$ and let a T-transformation of E_{n-1} into E_{n-1} and its associated T-function $f(a^{(k)})$ be fixed. The JPA of $a^{(0)}$ is called periodic if there exist non-negative integers

$$\ell \geq 0 ; \quad m \geq 1$$

$$\min L = \ell, \quad \min M = m \tag{3.1}$$

$$T^{m+v} = T^{v}. \qquad (v=\ell,\ell+1,\dots) \tag{3.2}$$

THEOREM 2. The JPA of $a^{(0)} \in E_{n-1}$ is periodic if and only if there exist a non-negative integer L and a natural number M such that, for $\min L = \ell$ and $\min M = m$

$$b^{(m+v)} = b^{(v)}. \qquad (v=\ell,\ell+1,\dots) \tag{3.3}$$

Proof. Necessity is obvious. For if (3.2) holds, then $a^{(m+v)} = a^{(v)}$, $a^{(m+v+1)} = a^{(v+1)}$ for $v \geq \ell$; it then follows, from

(2.1), that $b^{(m+v)} = b^{(v)}$ for $v = \ell, \ell+1, \ldots$. To prove sufficiency, suppose (3.3) holds. We shall now define Numbers $B_i^{(v)}$ $(i=0,\ldots,n-1;$ $v=0,1,\ldots)$ as follows.

$$B_i^{(v)} = \delta iv \qquad (i,v=0,1,\ldots,n-1)$$

$$B_i^{(v+n)} = \sum_{j=0}^{n-1} b_{\ell+v}^{(j)} B_i^{(v+j)}. \quad (b_{\ell+v}^{(0)}=1; i=0,\ldots,n-1; v=0,1,\ldots)$$

We then obtain, by Theorem 1, for $i=1,\ldots,n-1$

$$a_i^{(v)} = \lim_{j\to\infty} \frac{B_i^{(j)}}{B_0^{(j)}}, \qquad (v=\ell,\ell+1,\ldots)$$

but also

$$a_i^{(v+m)} = \lim_{j\to\infty} \frac{B_i^{(j)}}{B_0^{(j)}}, \qquad (v=\ell,\ell+1,\ldots)$$

so that indeed $a_i^{(v)} = a_i^{(v+m)}$ $(v=1,\ldots,n-1;\ v=\ell)$.

For periodic JPA we shall need the following

Definition VI. Let the JPA of $a^{(0)} \in E_{n-1}$ be periodic, viz. $a^{(m+v)} = a^{(v)}$ $(v=\ell,\ell+1,\ldots)$. The ℓ vectors

(3.4) $$b^{(0)}, b^{(1)}, \ldots, b^{(\ell-1)}$$

are said to form the primitive pre-period of this JPA, and the m vectors

(3.5) $$b^{(\ell)}, b^{(\ell+1)},\ldots,b^{(\ell+m-1)}$$

its primitive period. ℓ and m are called the length of the pre-period and the length of the period respectively. If $\ell = 0$, then the vectors $b^{(0)}, b^{(1)},\ldots,b^{(m-1)}$ form the primitive period of the JPA which is then called purely-periodic.

The concept of the primitive pre-period and the primitive period takes its origin from the following observation. Of course, we could take for the pre-period of the JPA any set of vectors

$$b^{(0)},\ b^{(1)},\ldots,b^{(\ell+sm-1)}$$

and for the period any set of vectors

$$b^{(\ell+sm)},b^{(\ell+sm+1)},\ldots,b^{(\ell+tm-1)},$$

with $t > s$, $s=0,1,\ldots$; here

$$\ell+sm-1 = L,\quad \ell+tm-1 = M$$

L and M from definition V; for s=0, t=1 we obtain the smallest lengths of the pre-period and the period, and these are then called primitive.

§ 2. The Function $f(a^{(k)}) = \left[a^{(k)}\right]$

The following notation will be used, in analogy with the greatest integer function

$$(3.6)\qquad \left[a^{(k)}\right] \equiv \left(\left[a_1^{(k)}\right],\left[a_2^{(k)}\right],\ldots,\left[a_{n-1}^{(k)}\right]\right). \qquad (k=0,1,\ldots)$$

Here, as is customary, for any real x, $[x] \in I$, and $[x] \leqslant x < [x]+1$. The T-function $f(a^{(k)}) = \left[a^{(k)}\right]$ occupies a very significant place in the theory of the JPA. Its main advantage rests with the fact that the $b^{(k)}$ are integral rational vectors (i.e. vectors with integral rational components). The rational integrality of the vectors $b^{(k)}$ is a most useful instrument for solving number-theoretic problems. For n=2, the JPA with the T-function $f(a_1^{(k)}) = \left[a_1^{(k)}\right]$ becomes the Euclidean Algorithm and yields the expansion of any real number by simple continued fraction. Besides, this T-function was originally used by Jacobi and Perron. That it also has its disadvantages, will be seen in the sequel. The vigor of the function $f(a^{(k)}) = \left[a^{(k)}\right]$ will be demonstrated by one of the following theorems of this paragraph. For this purpose we introduce the polynomial

$$(3.7)\qquad F(x) = x^n + k_1x^{n-1} + k_2x^{n-2} + \cdots + k_{n-1}x - d$$

with the restrictions on the coefficients

$$(3.8)\quad \begin{cases} \text{(i)}\ k_j\ (j=1,\dots,n-1),\ d \text{ are non-negative rational integers,} \\ \text{(ii)}\ k_j\ (j=1,\dots,n-2) \geqslant 0;\ k_{n-1},\ d \geqslant 1; \\ \text{(iii)}\ d \mid k_j\ ,\quad (j=1,\dots,n-1) \\ \text{(iv)}\ k_{n-1} \geqslant cd(n+k_1+k_2+\cdots+k_{n-2}),\ c \text{ any real number} > 1. \end{cases}$$

A polynomial (3.7) with coefficients from (3.8) will be called a P-polynomial of first order. Some of the most significant properties of the JPA are stated in

THEOREM 3. A P-polynomial $F(x)$ of first order has the following properties:

(i) $F(x)$ has one and only one real root w in the open interval

$$\left(0,\ \frac{1}{n+k_1+\cdots+k_{n-2}}\right);$$

(ii) The JPA with the T-function $f(a^{(k)}) = \left[a^{(k)}\right]$ of the vector

$$(3.9)\qquad a^{(0)}=(w+k_1,w^2+k_1w+k_2,w^3+k_1w^2+k_2w+k_3,\dots,w^{n-1}+k_1w^{n-2}+\cdots+k_{n-1})$$

is purely periodic; the length m of the primitive period equals

$$(3.10)\qquad m=n \text{ for } d>1;\quad m=1 \text{ for } d=1,$$

the primitive period has the structure

$$(3.11)\quad \begin{cases} b^{(0)} = (k_1,k_2,\dots,k_{n-1}), \\ b^{(i)} = (k_1,k_2,\dots,k_{n-1-i},k'_{n-i},k'_{n-i+1},\dots,k'_{n-1}),\ (i=1,\dots,n-2) \\ b^{(n-1)} = (k'_1,k'_2,\dots,k'_{n-1}), \\ k'_j = d^{-1}k_j,\qquad (j=1,\dots,n-1) \end{cases}$$

for $d > 1$; for $d = 1$, its structure is $b^{(0)}$;

(iii) $F(x)$ is irreducible over the field of rationals.

Proof. This follows the lines of the author's paper [2,a]. We first prove (i). Since $F(0) = -d < 0$, $F(1) = 1+k_1+\cdots+k_{n-1} - d > 0$ in virtue of (3.8), (IV), and since

$$F'(x) = k_{n-1} + \sum_{j=0}^{n-2}(n-j)k_jx^{n-j-1} > 0 \text{ for } x \geqslant 0$$

(here $k_0 = 1$), $F(x)$ has indeed one and only one root w in the open interval $(0,1)$. With $0 < w < 1$ and $w^n + k_1 w^{m-1} + \cdots + k_{n-2}w^2 + k_{n-1}w = d$, we obtain, in view of (3.8)

$$d > k_{n-1}w \geqslant d(n+k_1 + \cdots + k_{n-2})w,$$

$$w < \frac{1}{n+k_1 + \cdots + k_{n-2}},$$

which completes the proof of (i).

To prove (ii) we shall need the formula

$$\left[w^s+k_1w^{s-1}+\cdots+ k_{s-1}w+k_s\right] = k_s. \qquad (s=1,\ldots,n-1) \tag{3.12}$$

We must prove the inequalities

$$k_s \leqslant w^s + k_1w^{s-1} + \cdots + k_{s-1}w+k_s < k_s+1.$$

Since $w > 0$, we have

$$w^s+k_1w^{s-1} + \cdots + k_{s-1}w + k_s > k_s.$$

We thus have to prove

$$w^s+k_1w^{s-1} + \cdots + k_{s-1}w < 1.$$

But, in virtue of (i), we obtain easily

$$\begin{aligned} w^s+k_1w^{s-1} + \cdots + k_{s-1}w &= w(w^{s-1}+k_1w^{s-2} + \cdots + k_{s-2}w+k_{s-1}) \\ &< w(1+k_1 + \cdots + k_{s-2} + k_{s-1}) \\ &\leqslant w(1+k_1 + \cdots + k_{n-3} + k_{n-2}) \\ &< w(n+k_1 + \cdots + k_{n-3} + k_{n-2}) \\ &< \frac{1}{n+k_1 + \cdots + k_{n-2}}(n+k_1 + \cdots + k_{n-2}) = 1, \end{aligned}$$

which proves (3.12). It can be verified easily that, because of (3.12) and since $d \mid k_s$, the formula

$$\left[\frac{w^s+k_sw^{s-1} + \cdots + k_{s-1}w + k_s}{d}\right] = \frac{k_s}{d} \tag{3.13}$$

also holds.

We are now sufficiently equipped to attack the proof of (ii). We shall carry out the JPA of $a^{(0)}$ with the T-function $f(a^{(k)}) = \left[a^{(k)}\right]$. We obtain from (3.9), in virtue of (3.12)

(3.14) $$b^{(0)} = (k_1,k_2,\dots,k_{n-1}).$$

We shall repeatedly make use of the relations

(3.15) $$\frac{1}{w} = d^{-1}\,a_{n-1}^{(0)};$$

(3.16) $$a_s^{(0)} - b_s^{(0)} = wa_{s-1}^{(0)}. \qquad (s=1,\dots,n-1;\ a_0^{(0)} = 1)$$

Both formulas (3.15), (3.16) are easily verified; the first one is obtained from

$$w^n+k_1w^{n-1} +\dots+ k_{n-1}w = d,$$

$$\frac{1}{w} = \frac{w^{n-1}+k_1w^{n-2} +\dots+ k_{n-2}}{d} = d^{-1}\,a_{n-1}^{(0)};$$

the second one from

$$a_s^{(0)} - b_s^{(0)} = (w^s+k_1w^{s-1} +\dots+ k_{s-1}w + k_s) - k_s$$
$$= w(w^{s-1}+k_1w^{s-2} +\dots+ k_{s-1}) = wa_{s-1}^{(0)}.$$

We now obtain, from (3.15), (3.16)

$$a^{(1)} = \frac{1}{a_1^{(0)}-b_1^{(0)}}\,(a_2^{(0)}-b_2^{(0)},\dots,a_{n-1}^{(0)}-b_{n-1}^{(0)},\ 1)$$
$$= \frac{1}{w}\,(wa_1^{(0)},\dots,wa_{n-2}^{(0)},\ 1),$$

(3.17) $$a^{(1)} = (a_1^{(0)},a_2^{(0)},\dots,a_{n-1}^{(0)},\ d^{-1}\,a_{n-1}^{(0)}).$$

We now obtain from (3.17), in virtue of (3.12), (3.13),

(3.18) $$b^{(1)} = (k_1,k_2,\dots,k_{n-2},k'_{n-1}).$$

We recall the notation $k'_s = d^{-1}k_s$ $(s=1,\dots,n-1)$ and shall also use the notation

(3.19) $$a_s^{(0)'} = d^{-1}\, a_s^{(0)}. \qquad (s=1,\dots,n-1)$$

From (3.16) we then obtain $a_s^{(0)'} - b_s^{(0)'} = wa_{s-1}^{(0)'}$. We shall now prove the important formula

(3.20)
$$a^{(s)} = (a_1^{(0)}, a_2^{(0)}, \dots, a_{n-s-1}^{(0)}, a_{n-s}^{(0)'}, \dots, a_{n-1}^{(0)'}) \quad (s=1,\dots,n-2)$$
$$a^{(n-1)} = (a_1^{(0)'}, a_2^{(0)'}, \dots, a_{n-1}^{(0)'}).$$

We shall prove (3.20) by induction. It is correct for s=1, in virtue of (3.17). Let it be correct for s=t, viz.

(3.21) $$a^{(t)} = (a_1^{(0)}, a_2^{(0)}, \dots, a_{n-t-1}^{(0)}, a_{n-t}^{(0)'}, \dots, a_{n-1}^{(0)'}). \quad (1 \leqslant t \leqslant n-2)$$

We obtain from (3.21), in virtue of (3.12) and (3.13),

(3.22) $$b^{(t)} = (k_1, k_2, \dots, k_{n-t-1}, k'_{n-t}, \dots, k'_{n-1}). \qquad (1 \leqslant t \leqslant n-2)$$

We can also write (3.22) in the form

$$b^{(t)} = (b_1^{(0)}, b_2^{(0)}, \dots, b_{n-t-1}^{(0)}, b_{n-t}^{(0)'}, \dots, b_{n-1}^{(0)'}).$$

We now obtain from (3.21), (3.22)

$$\begin{aligned} a^{(t+1)} &= \frac{1}{a_1^{(0)} - b_1^{(0)}} \Big(a_2^{(0)} - b_2^{(0)}, \dots, a_{n-t-1}^{(0)} - b_{n-t-1}^{(0)}, \\ &\qquad a_{n-t}^{(0)'} - b_{n-t}^{(0)'}, \dots, a_{n-1}^{(0)'} - b_{n-1}^{(0)'}, 1 \Big) \\ &= \frac{1}{w} \Big(wa_1^{(0)}, \dots, wa_{n-t-2}^{(0)}, wa_{n-t-1}^{(0)'}, \dots, wa_{n-2}^{(0)'}, 1 \Big) \\ &= \Big(a_1^{(0)}, \dots, a_{n-t-2}^{(0)}, a_{n-t-1}^{(0)'}, \dots, a_{n-2}^{(0)'}, a_{n-1}^{(0)'} \Big) \end{aligned}$$

which proves formula (3.20).

From $a^{(n-1)} = (a_1^{(0)'}, a_2^{(0)'}, \dots, a_{n-1}^{(0)'})$ we again obtain, because of (3.13),

(3.23) $$b^{(n-1)} = \Big(b_1^{(0)'}, b_2^{(0)'}, \dots, b_{n-1}^{(0)'} \Big) = (k'_1, k'_2, \dots, k'_{n-1})$$

so that

$$a^{(n)} = \frac{1}{a_1^{(0)'}-b_1^{(0)'}}\left(a_2^{(0)'}-b_2^{(0)'},\ldots,a_{n-1}^{(0)'}-b_{n-1}^{(0)'}\right)$$

$$= \frac{d}{w}\left(wa_1^{(0)'},wa_2^{(0)'},\ldots,wa_{n-2}^{(0)'},\ 1\right)$$

$$= (da_1^{(0)'},da_2^{(0)'},\ldots,da_{n-2}^{(0)'},\ \tfrac{d}{w})$$

$$= (a_1^{(0)},\ a_2^{(0)},\ldots,a_{n-1}^{(0)}),$$

(3.24) $$a^{(n)} = a^{(0)}.$$

With (3.24) and (3.14), (3.22), (3.23), part (ii) of Theorem 3 is proved for $d > 1$. For $d = 1$, we obtain, from (3.18), $b^{(1)} = b^{(0)}$ and also $a^{(1)} = a^{(0)}$. This completes the proof of (ii).

To prove (iii) of Theorem 3 we shall need

Lemma 1. Under the conditions of Theorem 3, the sequences $\langle A_i^{(0)}-a_i^{(0)}A_0^{(0)}\rangle$ $(i=1,\ldots,n-1)$ are all null sequences.

Proof. We obtain, substituting for $a_i^{(0)}$ its value from (1.14),

$$A_i^{(v+n-1)} - a_i^{(0)}A_0^{(v+n-1)} = A_i^{(v+n-1)} - \frac{\left(\sum_{j=0}^{n-1} a_j^{(v)}A_i^{(v+j)}\right) A_0^{(v+n-1)}}{\sum_{j=0}^{n-1} a_j^{(v)}A_0^{(v+j)}}$$

$$= \frac{\sum_{j=0}^{n-1} A_i^{(v+n-1)}A_0^{(v+j)} - A_0^{(v+n-1)}A_i^{(v+j)} \quad a_j^{(v)}}{\sum_{j=0}^{n-1} a_j^{(v)}A_0^{(v+j)}};$$

$$\sum_{j=0}^{n-1} A_i^{(v+n-1)}A_0^{(v+j)}a_j^{(v)} - a_i^{(0)}\sum_{j=0}^{n-1} A_0^{(v+n-1)}A_0^{(v+j)}a_j^{(v)}$$

$$= \sum_{j=0}^{n-1} A_i^{(v+n-1)}A_0^{(v+j)}a_j^{(v)} - \sum_{j=0}^{n-1} A_0^{(v+n-1)}A_i^{(v+j)}a_j^{(v)};$$

after subtracting equal summands on both sides, and then cancelling by $A_0^{(v+n-1)} \neq 0$ for $v > 0$ we obtain

$$(3.25)\qquad -\left(A_i^{(v+n-1)} - a_i^{(0)} A_0^{(v+n-1)}\right) a_{n-1}^{(v)} = \sum_{j=0}^{n-2} \left(A_i^{(v+j)} - a_i^{(0)} A_0^{(v+j)}\right) a_j^{(v)} .$$

Denoting

$$(3.26)\qquad A_i^{(s)} - a_i^{(0)} A_0^{(s)} = H_i^{(s)} ,$$

we obtain from (3.25)

$$(3.27)\qquad -H_i^{(v+n-1)} = \frac{\sum_{j=0}^{n-2} H_i^{(v+j)} a_j^{(v)}}{a_{n-1}^{(v)}} ;$$

denoting further

$$(3.28)\qquad M_i^{(v)} = \max\left[\left|H_i^{(v)}\right|, \left|H_i^{(v+1)}\right|, \ldots, \left|H_i^{(v+n-2)}\right|\right]$$

and noting that $a_j^{(v)} > 0$ $(j=1,\ldots,n-1)$, we obtain from (3.27)

$$(3.29)\qquad \left|H_i^{(v+n-1)}\right| \leqslant \frac{M_i^{(v)} \sum_{j=0}^{n-2} a_j^{(v)}}{a_{n-1}^{(v)}} .$$

Since $\left[a_j^{(v)}\right] = k_j$ or k_j', we obtain, for $v \geqslant 0$,

$$a^{(v)} < k_j + 1;\ a_{n-1}^{(v)} \geqslant k_{n-1}' = d^{-1} k_{n-1},$$

so that, from (3.29)

$$H_i^{(v+n-1)} < \frac{M_i^{(v)} \sum_{j=0}^{n-2} (1+k_j)}{d^{-1} k_{n-1}} = \frac{M_i^{(v)} (n+k_i+\cdots+k_{n-2})}{d^{-1} k_{n-1}} ;$$

$$(3.30)\qquad \left|H_i^{(v+n-1)}\right| < \frac{1}{c} M_i^{(v)} .$$

Since

$$M_i^{(v+1)} = \max\left[\left|H_i^{(v+1)}\right|, \left|H_i^{(v+2)}\right|, \ldots, \left|H_i^{(v+n-2)}\right|, \left|H_i^{(v+n-1)}\right|\right]$$

$$\leqslant \max\left[M_i^{(v)}, \frac{1}{c} M_i^{(v)}\right] = M_i^{(v)} ,$$

we have obtained

(3.31) $$M_i^{(v+1)} \leqslant M_i^{(v)}.$$

From (3.30), (3.31) we now obtain

$$\left|H_i^{(v+n)}\right| < \frac{1}{c} M_i^{(v+1)} \leqslant \frac{1}{c} M_i^{(v)},$$

$$\left|H_i^{(v+n-1)}\right| < \frac{1}{c} M_i^{(v+2)} \leqslant \frac{1}{c} M_i^{(v)},$$

- - - - - - - - - - - - - - - -

$$H_i^{|v+n+n-2|} < \frac{1}{c} M_i^{(v+n-1)} \leqslant \frac{1}{c} M_i^{(v)},$$

so that, since $M_i^{(v+n)} = \left[\left|H_i^{(v+n)}\right|, \left|H_i^{(v+n+1)}\right|, \ldots, \left|H_i^{(v+2n-2)}\right|\right]$

(3.32) $$M_i^{(v+n)} < \frac{1}{c} M_i^{(v)}.$$

Let $v = sn+t$, $0 \leqslant t \leqslant n-1$, then, from (3.32),

$$M_i^{(v+n)} = M_i^{((s+1)n+t)} < \frac{1}{c} M_i^{(sn+t)} < \frac{1}{c^2} M_i^{((s-1)n+t)}$$

$$< \frac{1}{c^3} M_i^{((s-2)n+t)} < \cdots < \frac{1}{c^{s+1}} M_i^{t} \leqslant \frac{1}{c^{s+1}} M_i^{(0)},$$

(3.33) $$M_i^{((s+1)n+t)} < \frac{1}{c^{s+1}} M_i^{(0)}. \quad (s=0,1,\ldots;\ 0 \leqslant t \leqslant n-1)$$

But

$$M_i^{(0)} = \max\left[\left|A_i^{(0)} - a_i^{(0)} A_0^{(0)}\right|, \left|A_i^{(1)} - a_i^{(0)} A_0^{(1)}\right|, \ldots,\right.$$

$$\left.\left|A_i^{(i)} - a_i^{(0)} A_0^{(i)}\right|, \ldots, \left|A_i^{(n-2)} - a_i^{(0)} A_0^{(n-2)}\right|\right]$$

$$= \max\left[\left|-a_i^{(0)}\right|, 0, \ldots, 0, 1, 0, \ldots, 0\right]$$

$$= \max\left[\left|a_i^{(0)}\right|, 1\right] \leqslant 1 + k_i,$$

so that, from (3.33)

$$M_i^{((s+1)n+t)} < \frac{1 + k_i}{c^{s+1}}.$$

Denoting $(s+1)n+t = v$, $s+1 = \frac{v-t}{n} \geqslant \frac{v}{n} \geqslant \left[\frac{v}{n}\right]$, we obtain from (3.33)

$$(3.34)\qquad M_i^{(v)} < \frac{1 + k_i}{c^{\left[\frac{v}{n}\right]}} .$$

Since $c > 1$, (3.34) shows that $\langle M_i^{(v)} \rangle$ is a null sequence, and with it also $\langle H_i^{(v)} \rangle = \langle A_i^{(v)} - a_i^{(0)} A_0^{(v)} \rangle$ is a null sequence. Lemma 1 is proved.

We shall now show that $F(x)$ or $F(w)$ is irreducible over the field of rationals. Presuming this is not the case, there exists an equation of the type

$$(3.35)\qquad c_0 w^m + c_1 w^{m-1} + \cdots + c_{m-1} w + c_m = 0,$$

$$c_i \in I\ (i=0,\ldots,m);\ 1 \leqslant m \leqslant n-1.\ c_m \neq 0;\ c_0 \neq 0.$$

From (3.9) we obtain

$$a_1^{(0)} = w + k_1;\quad w = a_1^{(0)} - k_1$$

$$a_2^{(0)} = w^2 + k_1 w + k_2;\quad w^2 = a_2^{(0)} - k_1(a_1^{(0)} - k_1) - k_2.$$

$$w^2 = a_2^{(0)} - k_1 a_1^{(0)} + (k_1^2 - k_2)$$

$$a_3^{(0)} = w^3 + k_1 w^2 + k_2 w + k_3;$$

$$w^3 = a_3^{(0)} - k_1 \left[a_2^{(0)} - k_1 a_1^{(0)} + (k_1^2 - k_2)\right] - k_2(a_1^{(0)} - k_1) - k_3,$$

$$w^3 = a_3^{(0)} - k_1 a_2^{(0)} + (k_1^2 - k_2) a_1^{(0)} - (k_1^3 - 2k_1 k_2 + k_3).$$

We shall prove the formula

$$(3.36)\qquad \begin{aligned} w^t &= \sum_{j=0}^{t} (-1)^j p_j^{(t)} a_{t-j}^{(0)}; \\ \sum_{j=0}^{s} (-1)^j k_j p_{s-j}^{(t)} &= 0. \qquad (p_0^{(t)} = k_0 = 1;\ s=1,2,\ldots;\ t=1,\ldots,m) \end{aligned}$$

The numbers $p_j^{(i)}$ are so-called partition polynomials which will be dealt with more extensively in a later chapter. The $p_j^{(i)}$ are all rational integers, since they are rational polynomials in the variables $k_1,\ldots,k_i$ with integral coefficients.

To prove (3.36) we observe that formula (3.36) is correct for $i=1,2,3$, then we use induction.

Substituting the values of w^i in (3.35) we obtain

$$(3.37)\quad \begin{gathered} p_0 a_m^{(0)} + p_1 a_{m-1}^{(0)} + p_2 a_{m-2}^{(0)} + \cdots + p_{m-1} a_1^{(0)} + p_m = 0, \\ p_0 = c_0;\ p_k = \sum_{j=0}^{k} (-1)^j c_{k-j} p_j^{(m-j)}. \quad (p_0^{(m)} = 1;\ k=0,\ldots,m). \end{gathered}$$

We shall now presume that all coefficients p_i $(i=1,\ldots,n)$ but the first vanish, viz. $p_i = 0$; $(i=1,\ldots,m)$ it then follows (as can be easily verified by the reader) from (3.37), viz.
$p_k = \sum_{j=0}^{k} (-1)^j c_{k-j} p_j^{(m-j)}$, that in this case

$$(3.38)\qquad c_i = c_0 k_i. \quad (i=1,2,\ldots,n)$$

Substituting the values of c_i from (3.38) in (3.35) and cancelling by $c_0 \neq 0$, we obtain

$$(3.39)\qquad w^m + k_1 w^{m-1} + k_2 w^{m-2} + \cdots + k_{m-1} w + k_m = 0.$$

But equation (3.39) cannot hold; since $k_i \geqslant 0$ $(i=1,\ldots,m)$ we would obtain $w \leqslant 0$, contrary to $0 < w < \frac{1}{n+k_1+\cdots+k_{n-2}}$. Thus $p_i = 0$ cannot hold for all i, and we can presume, without loss of generality

$$(3.40)\qquad p_{m-1} \neq 0.$$

Now, according to (1.15)

$$\frac{\pm 1}{A_0^{(v)} + a_1^{(v)} A_0^{(v+1)} + \cdots + a_{n-1}^{(v)} A_0^{(v+n-1)}} = \begin{vmatrix} 1 & A_0^{(v+1)} & \cdots & A_0^{(v+n-1)} \\ a_1^{(0)} & A_1^{(v+1)} & \cdots & A_1^{(v+n-1)} \\ a_2^{(0)} & A_2^{(v+1)} & \cdots & A_2^{(v+n-1)} \\ \vdots & \vdots & & \vdots \\ a_{n-1}^{(0)} & A_{n-1}^{(v+1)} & \cdots & A_{n-1}^{(v+n-1)} \end{vmatrix}$$

and, subtracting from the i-th row vector the $a_i^{(0)}$-multiple of the first row vector $(i=2,\dots,n-1)$

$$\frac{\pm 1}{\sum_{j=0}^{n-1} a_j^{(v)} A_0^{(v+j)}} =$$

(3.41) $$\begin{vmatrix} A_1^{(v+1)}-a_1^{(0)}A_0^{(v+1)}, & A_1^{(v+2)}-a_1^{(0)}A_0^{(v+2)}, & \dots, & A_1^{(v+n-1)}-a_1^{(0)}A_0^{(v+n-1)} \\ A_2^{(v+1)}-a_2^{(0)}A_0^{(v+1)}, & A_2^{(v+2)}-a_2^{(0)}A_0^{(v+2)}, & \dots, & A_2^{(v+n-1)}-a_2^{(0)}A_0^{(v+n-1)} \\ \vdots & \vdots & & \vdots \\ A_{n-1}^{(v+1)}-a_{n-1}^{(0)}A_0^{(v+1)}, & A_{n-1}^{(v+2)}-a_{n-1}^{(0)}A_0^{(v+2)}, & \dots, & A_{n-1}^{(v+n-1)}-a_{n-1}^{(0)}A_0^{(v+n-1)} \end{vmatrix}.$$

Substituting in (3.37)

(3.42) $$Q_i = p_{m-i} \qquad (i=0,1,\dots,n)$$

this equation takes the form

(3.43) $$Q_0 + a_1^{(0)}Q_1 + a_2^{(0)}Q_2 + \dots + a_{n-1}^{(0)}Q_{n-1} = 0,$$

where we have to set $Q_j = 0$, if $j > m$. Multiplying both sides of (3.41) by $Q_1 = p_{m-1} \neq 0$, this equation becomes

(3.44) $$\frac{\pm Q_1}{\sum_0^{n-1} a_j^{(v)} A_0^{(v+j)}} =$$

$$\begin{vmatrix} Q_1(A_1^{(v+1)}-a_1^{(0)}A_0^{(v+1)}), & Q_1(A_1^{(v+2)}-a_1^{(0)}A_0^{(v+2)}), & \dots, & Q_1(A_1^{(v+n-1)}-a_1^{(0)}A_0^{(v+n-1)}) \\ A_2^{(v+1)}-a_2^{(0)}A_0^{(v+1)} & ,A_2^{(v+2)}-a_2^{(0)}A_0^{(v+2)} & ,\dots, & A_2^{(v+n-1)}-a_2^{(0)}A_0^{(v+n-1)} \\ \vdots & \vdots & & \vdots \\ A_{n-1}^{(v+1)}-a_{n-1}^{(0)}A_0^{(v+1)} & ,A_{n-1}^{(v+2)}-a_{n-1}^{(0)}A_0^{(v+2)} & ,\dots, & A_{n-1}^{(v+n-1)}-a_{n-1}^{(0)}A_0^{(v+n-1)} \end{vmatrix}.$$

If we add to the first row vector of the determinant in (3.44) the Q_2-multiple of the second row vector the Q_3-multiple of the third row vector,..., the Q_{n-1}-multiple of the n-1-th row vector, we obtain that the elements c_{1k} of the first row now take the form

(3.45) $$c_{1,k} = \sum_{j=0}^{n-2} Q_{j+1}\left(A_{j+1}^{(v+k)} - a_{j+1}^{(0)}A_0^{(v+k)}\right). \qquad (k=1,\dots,n-1)$$

But, by Lemma 1, $\left\langle A_{j+1}^{(v+k)} - a_{j+1}^{(0)} A_0^{(v+k)} \right\rangle$, is a null sequence; let

$$(3.46) \qquad \left| A_{j+1}^{(v+k)} - a_{j+1}^{(0)} A_0^{(v+k)} \right| < \varepsilon_k;$$

From (3.45), (3.46) we obtain $\left| c_{1,k} \right| \leqslant \varepsilon_k \sum_{j=0}^{n-2} \left| Q_{j+1} \right|$, and choosing

$$\varepsilon_k < \frac{1}{\sum_{j=1}^{n-2} \left| Q_{j+1} \right|},$$

we obtain

$$(3.47) \qquad \left| c_{1,k} \right| < 1.$$

But $c_{1,k}$ is an integer, we therefore conclude

$$(3.48) \qquad c_{1,k} = 0. \qquad (k=1,\ldots,n-1).$$

Because of (3.48) the determinant in the right side of (3.44) is zero, while the left side is never zero, since $Q_1 \neq 0$, $\sum_{j=0}^{n-1} a_j A_0^{(v+j)} < \infty$. This contradiction is a result of our presumption that F(w) is reducible. **Thus Theorem 3 is completely proved .**

<u>Definition VII.</u> A JPA of $a^{(0)} \in E_{n-1}$ such that the sequences

$$\left\langle A_i^{(v)} - a_i^{(0)} A_0^{(v)} \right\rangle \qquad (i=1,\ldots,n-1)$$

are all null sequences is said to be <u>ideally convergent</u>.

A JPA which is ideally convergent is also convergent, if only

$$\left| A_0^{(v)} \right| > 1, \qquad \text{for } v > v_0.$$

For it follows from $\left| A_i^{(v)} - a_i^{(0)} A_0^{(v)} \right| < \varepsilon$ for $v > v_0(\varepsilon)$

$$\left| a_i^{(0)} - \frac{A_i^{(v)}}{A_0^{(v)}} \right| < \frac{\varepsilon}{\left| A_0^{(v)} \right|} < \varepsilon.$$

If the JPA of $a^{(0)} \in E_{n-1}$ is associated with the T-function $f(a^{(k)}) = \left[a^{(k)} \right]$, then $A_0^{(v)} > 1$ for any $v \geqslant n$ so that ideal convergence here always implies convergence. But, as the reader can easily verify, the JPA of $a^{(0)} \in E_{n-1}$ with the associated T-function $f(a^{(k)}) = \left[a^{(k)}\right]$ is always convergent, since in this case

$$a_{n-1}^{(v)} = \frac{1}{a_1^{(v-1)} - b_1^{(v-1)}} > 1, \qquad (v \geqslant 1)$$

so that $b_{n-1}^{(v)} \geqslant 1$, and it is also easily verified that

$$0 \leqslant \frac{b_i^{(v)}}{b_{n-1}^{(v)}} < 1. \qquad (v = 1, 2, \ldots)$$

But we can also always achieve that $b_{n-1}^{(0)} \geqslant 1$; for if $\left[a_{n-1}^{(0)}\right] = -m < 0$, then, substituting $a_{n-1}^{(0)'} = a_{n-1}^{(0)} + m + 1$, we have $b_{n-1}^{(v)'} = 1$; the same holds for any $a_i^{(0)}$ $(i=1,\ldots,n-2)$. Since the JPA of $a^{(0)} \in E_{n-1}$ with the T-function $f(a^{(k)}) = \left[a^{(k)}\right]$ is convergent, we obtain, for a rational approximation of the $a_i^{(0)}$

$$a^{(0)} = \lim_{v \to \infty} \frac{A_i^{(v)}}{A_0^{(v)}}. \qquad (i=1,\ldots,n-1)$$

If, as before $a^{(0)} = (w+k_1, w^2+k_1w+k_2, \ldots, w^{n-1}+k_1w^{n-2}+\cdots+k_{n-1})$ then we wish to approximate w, and we obtain from the limit formula, for $i = 1$,

$$w + k_1 = \lim_{v \to \infty} \frac{A_1^{(v)}}{A_0^{(v)}}.$$

We shall prove the formula

$$(3.49) \qquad A_1^{(sn+1)} = k_1 A_0^{(sn+1)} + A_0^{(sn)}. \qquad (s=0,1,\ldots)$$

Proof is by induction; the following n identities are easily verified

$$(3.50) \qquad \begin{aligned} A_1^{(j)} &= k_1 A_0^{(j)} + A_0^{(j-1)}, \qquad (j=1,\ldots,n-1) \\ A_1^{(n)} &= k_1 A_0^{(n)} + dA_0^{(n-1)}. \end{aligned}$$

The last equation of (3.50) is correct, since $A_1^{(n)} = k_1$, $A_0^{(n-1)} = 0$. We presume the correctness of the following n identities

$$(3.51)\qquad \begin{aligned} A_1^{(tn+j)} &= k_1A_0^{(tn+j)} + A_0^{(tn+j-1)}, \quad (j=1,\dots,n-1)\\ A_1^{(t+1)n} &= k_1A_0^{(t+1)n} + dA_0^{(tn+n-1)}. \end{aligned}$$

We obtain, on the basis of (3.51),

$$\begin{aligned} A_1^{((t+1)n+1)} &= \left(\sum_{j=0}^{n-2} k_jA_1^{(tn+1+j)}\right) + k'_{n-1}A_1^{((t+1)n)}\\ &= \sum_{j=0}^{n-2} k_j\left(k_1A_0^{(tn+1+j)} + A_0^{(tn+j)}\right) + k'_{n-1}\left(k_1A_0^{((t+1)n)} + dA_0^{(tn+n-1)}\right)\\ &= k_1\left(\sum_{j=0}^{n-2} k_jA_0^{(tn+1+j)} + k'_{n-1}A_0^{((t+1)n)}\right) + \sum_{j=0}^{n-2} k_jA_0^{(tn+j)} + k_{n-1}A_0^{(tn+n-1)}\\ &= k_1A_0^{((t+1)n+1)} + A_0^{((t+1)n)}. \end{aligned}$$

In exactly the same way we prove the remaining n-1 identities, so that indeed

$$(3.52)\qquad \begin{aligned} A_1^{((t+1)n+j)} &= k_1A_0^{((t+1)n+j)} + A_0^{((t+1)n+j-1)}, \quad (j=1,\dots,n-1)\\ A_1^{((t+2)n)} &= k_1A_0^{((t+2)n)} + dA_0^{((t+1)n+n-1)}. \end{aligned}$$

With (3.50) and (3.52) formula (3.49) is completely proved. On the basis of this formula we now obtain from the limit relation

$$w + k_1 = \lim_{s\to\infty} \frac{A_1^{(sn+1)}}{A_0^{(sn+1)}} = \lim_{s\to\infty} \frac{k_1A_0^{(sn+1)} + A_0^{(sn)}}{A_0^{(sn+1)}} = k_1 + \lim \frac{A_0^{(sn)}}{A_0^{(sn+1)}},$$

$$(3.53)\qquad w = \lim_{s\to\infty} \frac{A_0^{(sn)}}{A_0^{(sn+1)}}.$$

For practical purposes formula (3.53) is very convenient; first, only the $A_0^{(v)}$ have to be calculated; second, the convergents of w are fractions where numerators and denominators are positive integers.

To obtain a measure of convergence for w, we return to formula (3.34) which results in

$$\left| A_1^{(v)} - a_1^{(0)} A_0^{(v)} \right| < \frac{1 + k_1}{c^{\left[\frac{v}{n}\right]}} ,$$

so that

$$\left| a_1^{(0)} - \frac{A_1^{(v)}}{A_0^{(v)}} \right| < \frac{1 + k_1}{c^{\left[\frac{v}{n}\right]} A_0^{(v)}} ,$$

and substituting $v = (s+1)n+1$,

$$\left| w + k_1 - \frac{k_1 A_0^{((s+1)n+1)} + A_0^{((s+1)n)}}{A_0^{((s+1)n+1)}} \right| < \frac{1 + k_1}{c^{s+1} A_0^{((s+1)n)}}$$

since $A_0^{((s+1)n+1)} > A_0^{((s+1)n)}$; thus

$$\left| w - \frac{A_0^{((s+1)n)}}{A_0^{((s+1)n+1)}} \right| < \frac{1 + k_1}{c^{s+1} A_0^{((s+1)n)}} . \tag{3.54}$$

We shall estimate $A_0^{((s+1)n)}$. We shall make use of the relation

$$\begin{aligned} A_0^{(v+1)} &\geq k'_{n-1} A_0^{(v)}, \qquad (v+1 \not\equiv 0 \pmod n) \\ A_0^{(v+1)} &\geq k'_{n-1} A_0^{(v)}. \qquad (v+1 \equiv 0 \pmod n) \end{aligned} \tag{3.55}$$

It follows from (3.55)

$$\begin{aligned} A_0^{(n+1)} &\geq k'_{n-1}, \\ A_0^{(n+2)} &\geq k'_{n-1} A_0^{(n+1)} \geq k'^{\,2}_{n-1}, \\ &- - - - - - - - - - - - - \\ A_0^{(n+n-1)} &\geq k'^{\,n-1}_{n-1}, \\ A_0^{(2n)} &\geq d k'^{\,(n)}_{n-1}, \end{aligned}$$

and generally

$$A_0^{((s+1)n)} \geq d^s k'^{\,sn}_{n-1} , \qquad (s=1,2,\dots)$$

or, since $k'_{n-1} = d^{-1} k_{n-1}$, $k_{n-1} \geq cd(n+k_1+\dots+k_{n-2})$

$$A_0^{(s+1)n} \geq d^s c^{sn} (n+k_1+\dots+k_{n-2})^{sn},$$

so that, from (3.54)

$$(3.56) \qquad \left| w - \frac{A_0^{((s+1)n)}}{A_0^{((s+1)n)}} \right| < \frac{1 + k_1}{c^{sn+s+1} d^s (n+k_1+\cdots+k_{n-2})^{sn}} .$$

A less exact, but simple approximation measure would be given by

$$(3.57) \qquad \left| w - \frac{A_0^{((s+1)n)}}{A_0^{((s+1)n+1)}} \right| < \frac{1 + k_1}{((nc)^n d)^s} .$$

(3.57) shows that in the case of ideal convergence, the convergence is indeed ideal.

Example 6. Let

$$F(x) = x^5 + 2x^3 + 2x^2 + 28x - 2.$$

Here $k_1 = 0$, $k_2 = 2$, $k_3 = 2$, $k_4 = 28$; $d = 2$; $d | k_i$. Since $28 > \frac{3}{2} \cdot 2(5 + 2 + 2)$, $(c = \frac{3}{2})$ the conditions of Theorem 3 are fulfilled. We have

$$0 < w < \frac{1}{9};$$

w is a fifth degree (algebraic) irrational; the JPA with $f(a^{(k)}) = \left[a^{(k)}\right]$ of $a^{(0)} = (w, w^2+2, w^3+2w+2, w^4+2w^2+2w+28)$ has the primitive period

$$b^{(0)} = (0, 2, 2, 28),$$
$$b^{(1)} = (0, 2, 2, 14),$$
$$b^{(2)} = (0, 2, 1, 14),$$
$$b^{(3)} = (0, 1, 1, 14),$$
$$b^{(4)} = (0, 1, 1, 14).$$

One calculates easily

$$A_0^{(10)} = 1{\cdot}098{\cdot}865$$

$$A_0^{(11)} = 15{\cdot}467{\cdot}736.$$

We now obtain from (3.57), for s=1,

$$\left| w - \frac{1\cdot 098\cdot 865}{15\cdot 467\cdot 736} \right| < \frac{1}{(5\cdot\frac{3}{2})^{5}\cdot 2} < .000022$$

$$w = .0710 \pm .000022 .$$

Chapter 4.

SOME SPECIAL CASES OF JPA

In this chapter we shall enumerate a few special cases of periodic JPA's with $f(a^{(k)}) = \left[a^{(k)}\right]$. To do justice to history, these are the classical cases which were first investigated by C. G. J. Jacobi, P. Bachman, O. Perron and H. Hasse, and later by the author.

§1. Periodicity of $a^{(0)} \in E_{n-1}$, $\alpha = (D^n + d)^{\frac{1}{n}}$

Our starting point will again be the polynomial

$$F(x) = x^n+k_1x^{n-1}+k_2x^{n-2}+\cdots+k_{n-1}x - d,$$

and we shall now specify the coefficients by

$$k_i = \binom{n}{i}D^i;\quad d|D;\quad D\in N. \quad (i=1,\ldots,n-1) \tag{4.1}$$

We have, as before, $F(w) = 0$,

$$0 < w < \frac{1}{n+k_1+k_2+\cdots+k_{n-2}} .$$

If we now put

$$w = \alpha - D, \tag{4.2}$$

the equation $F(w) = 0$ becomes

$$(\alpha-D)^n + \binom{n}{1}D(\alpha-D)^{n-1} + \binom{n}{2}D^2(\alpha-D)^{n-2}+\cdots+ \binom{n}{n-1}D^{n-1}(\alpha-D) - d = 0$$

or

$$\left[(\alpha-D)+D\right]^n - D^n - d = 0$$

$$\alpha^n = D^n + d,$$

and, for the positive real root of α,

$$\alpha = \sqrt[n]{D^n + d} . \tag{4.3}$$

We shall now express the vector

$$a^{(0)} = (w+k_1,w^2+k_1w+k_2,\ldots,w^{n-1}+k_1w^{n-2}+\cdots+k_{n-2}w+k_{n-1})$$

as a function of α; we obtain for the s-th component of $a^{(0)}$ with

$k_0 = 1$

$$a_s^{(0)} = \sum_{i=0}^{s} k_i w^{s-i} = \sum_{i=0}^{j} \binom{n}{i} D^i (\alpha - D)^{s-i}.$$

Making use of the formula

$$\sum_{i=0}^{j} (-1)^i \binom{n}{j-i} \binom{s-j+i}{i} = \binom{n-s+j-1}{j}$$

$a_s^{(0)}$ takes the form, after easy rearrangements

$$(4.4) \qquad a_s^{(0)} = \sum_{i=0}^{s} \binom{n-s-1+i}{i} D^i \alpha^{s-i}. \qquad (s=1,\dots,n-1)$$

The choice of D,d is not free; rather they are functionally connected. We shall investigate the condition

$$k_{n-1} \geqslant cd(n+k_1+k_2+\dots+k_{n-2})$$

for the specified values of the k_i from (4.1); the reader may be reminded that the fulfillment of this condition guarantees that (i) the JPA of $a^{(0)} \in Q(\alpha)$, with components $a_s^{(0)}$ $(s=1,\dots,n-1)$ from (4.4) and the T-function $f(a^{(k)}) = \left[a^{(k)}\right]$ becomes periodic; (ii) this JPA is ideally convergent. The reader should note that ideal convergence of the JPA with $f(a^{(k)}) = \left[a^{(k)}\right]$ results in periodicity of $a^{(0)}$, as specified by (4.4), but not vice versa. The condition for ideal convergence now takes the form

$$(4.5) \qquad \binom{n}{1} D^{n-1} \geqslant cd\left(n + \binom{n}{1} D + \binom{n}{2} D^2 + \dots + \binom{n}{n-2} D^{n-2}\right).$$

We shall presume

$$(4.6) \qquad D \geqslant d(n-2) \geqslant (n-2), \qquad (n \geqslant 3)$$

and obtain

$$\binom{n}{n-2}D^{n-2}+\binom{n}{n-3}D^{n-3}+\cdots+\binom{n}{2}D^2+\binom{n}{1}D + n$$

$$= \binom{n}{2}D^{n-2}+\binom{n}{3}D^{n-3}+\cdots+\binom{n}{n-2}D^2+\binom{n}{n-1}D + n$$

$$= D^{n-2}\left[\binom{n}{2}+\binom{n}{3}\cdot\frac{1}{D}+\binom{n}{4}\cdot\frac{1}{D^2}+\cdots+\binom{n}{n-2}\cdot\frac{1}{D^{n-4}}+\binom{n}{n-1}\cdot\frac{1}{D^{n-3}}+\frac{n}{D^{n-2}}\right]$$

$$\leqslant D^{n-2}\left[\binom{n}{2}+\binom{n}{3}\cdot\frac{1}{n-2}+\binom{n}{4}\cdot\frac{1}{(n-2)^2}+\cdots\right.$$

$$\left.+\binom{n}{n-2}\cdot\frac{1}{(n-2)^{n-4}}+\binom{n}{n-1}\cdot\frac{1}{(n-2)^{n-3}}+\frac{n}{(n-2)^{n-2}}\right]$$

$$= D^{n-2}\left[\binom{n}{2}+\binom{n}{2}\frac{n-2}{3(n-2)}+\binom{n}{2}\frac{(n-2)(n-3)}{3\cdot 4(n-2)^2}+\cdots\right.$$

$$\left.+\binom{n}{2}\frac{(n-2)(n-3)\ldots 2}{3\cdot 4\ldots(n-1)(n-2)^{n-3}}+\frac{n}{(n-2)^{n-2}}\right]$$

$$\leqslant D^{n-2}\left[\binom{n}{2}+\binom{n}{2}\cdot\frac{1}{3}+\binom{n}{2}\cdot\frac{1}{3\cdot 4}+\cdots+\binom{n}{2}\cdot\frac{1}{3\cdot 4\ldots(n-1)}+n\right]$$

$$= \binom{n}{2}D^{n-2}\left[1+\frac{1}{3}+\frac{1}{3\cdot 4}+\cdots+\frac{1}{3\cdot 4\ldots(n-1)}+\frac{2}{n-1}\right]$$

$$\leqslant \binom{n}{2}D^{n-2}\left[1+\frac{1}{3}+\frac{1}{3\cdot 4}+\cdots+\frac{1}{3\cdot 4\ldots(n-1)}+1\right]$$

$$< \binom{n}{2}D^{n-2}\left[2+2\left(\frac{1}{2\cdot 3}+\frac{1}{2\cdot 3\cdot 4}+\frac{1}{2\cdot 3\cdot 4\cdot 5}+\cdots\right)\right]$$

$$= \binom{n}{2}D^{n-2}\left[2+2(e-2-\tfrac{1}{2})\right] = (e-\tfrac{3}{2})\, n(n-1)D^{n-2}.$$

Choosing

$$\binom{n}{1}D^{n-1} \geqslant cd(e-\tfrac{3}{2})n(n-1)D^{n-2} > cd\left[n+\binom{n}{1}D+\binom{n}{2}D^2+\cdots+\binom{n}{n-2}D^{n-2}\right]$$

we obtain

$$D \geqslant cd(e-\tfrac{3}{2})\ (n-1).$$

Putting $c = \dfrac{e}{2(e-\frac{3}{2})} > 1$, we obtain

$$D \geqslant \frac{e}{2}\,(n-1)\,d. \tag{4.7}$$

We had presumed $D \geqslant d(n-2)$; since $\frac{e}{2}(n-1) > n-2$, the result (4.7) is consistent with our presumption.

We shall now investigate the conditions under which the JPA of

$a^{(0)} \in E_{n-1}$; $a_s^{(0)} = \sum_{i=0}^{s} \binom{n-s-1+i}{i} D^i \alpha^{s-i}$, $\alpha = \sqrt[n]{D^n + d}$ with

$f(a^{(k)}) = \left[a^{(k)}\right]$ is periodic. From the proof of Theorem 3, it is obvious that periodicity occurs if

$$0 < w^s + k_1 w^{s-1} + \cdots + k_{s-2} w^2 + k_{s-1} w < 1;$$

but

$$w^s + k_1 w^{s-1} + \cdots + k_{s-2} w^2 + k_{s-1} w \leqslant w + k_1 w + \cdots + k_{s-1} w < 1$$

necessitates $w < \dfrac{1}{1 + k_1 + \cdots + k_{s-1}}$. Putting

$$w < \frac{1}{1 + k_1 + \cdots + k_{n-2}} \leqslant 1 \tag{4.9}$$

we see that the condition for periodicity is always satisfied, since $s \leqslant n-1$. We shall now investigate the condition (4.9) for the coefficients k_i as specified by (4.1). We obtain

$$w = \alpha - D = \sqrt[n]{D^n + d} - D = D\left(1 + \frac{d}{D^n}\right)^{\frac{1}{n}} - D$$

$$> D\left(1 + \frac{d}{nD^n}\right) - D, \qquad w > \frac{d}{nD^{n-1}},$$

and condition (4.9) becomes

$$\frac{d}{nD^{n-1}} < \frac{1}{1+\binom{n}{1}D+\binom{n}{2}D^2+\cdots+\binom{n}{n-2}D^{n-2}}.$$

Presuming, as before, $D \geqslant d(n-2)$, we obtain

$$1+\binom{n}{1}D+\binom{n}{2}D^2+\cdots+\binom{n}{n-2}D^{n-2} \leqslant$$

$$D^{n-2}\left[\binom{n}{2}+\binom{n}{3}\cdot\frac{1}{n-2}+\binom{n}{4}\cdot\frac{1}{(n-2)^2}+\cdots+\binom{n}{n}\frac{1}{(n-2)^{n-2}}\right]$$

$$< \binom{n}{2}D^{n-2}\left[1 + \frac{1}{3} + \frac{1}{3\cdot 4} + \frac{1}{3\cdot 4\cdot 5} + \cdots\right] =$$

$$\binom{n}{2}D^{n-2}\left[1 + 2\left(e - \frac{5}{2}\right)\right] = n(n-1)(e-2)D^{n-2}.$$

Putting $\frac{d}{nD^{n-1}} \leqslant \frac{1}{n(n-1)(e-2)D^{n-2}} < \frac{1}{1+\binom{n}{1}D+\binom{n}{2}D^2+\cdots+\binom{n}{n-2}D^{n-2}}$ we obtain

$$D \geqslant (n-1)(e-2)d. \tag{4.10}$$

Since this result has to be consistent with the presumed value of D, and since $(e-2)(n-1) > n-2$ for $n \leqslant 4$, and smaller otherwise, we obtain

$$\begin{aligned} D &\geqslant (e-2)(n-1)d, && n \leqslant 4; \\ D &\geqslant (n-2)d, && n > 4. \end{aligned} \tag{4.11}$$

For the case $n \leqslant 4$, a stronger result was obtained by the author in a previous paper [2,b].

We can now summarize these investigations in

THEOREM 4. Let α be the irrational

$$\alpha = (D^n+d)^{\frac{1}{n}}; \quad d,D \in N; \quad d|D;$$

$$D \geqslant \tfrac{13}{18}(n-1)d \text{ for } n \leqslant 4; \quad D \geqslant (n-2)d \quad \text{for } n > 4.$$

The JPA, with $f(a^{(k)}) = [a^{(k)}]$ of

$$a^{(0)} = \left(\sum_{i=0}^{1}\binom{n-2+i}{i}D^i\alpha^{1-i},\ \sum_{i=0}^{2}\binom{n-3+i}{i}D^i\alpha^{2-i},\ldots,\sum_{i=0}^{n-1}D^i\alpha^{n-1-i}\right)$$

is purely periodic with length of period $m = n$ for $d \neq 1$; $m = 1$ for $d = 1$. The period has the form

$$b^{(0)} = \left(\binom{n}{1}D,\ \binom{n}{2}D^2,\ldots,\binom{n}{n-1}D^{n-1}\right),$$

$$b^{(s)} = \left(\binom{n}{1}D,\ldots,\binom{n}{n-s-1}D^{n-s-1},\binom{n}{n-s}D^{n-s-1}t,\ldots,\binom{n}{n-1}D^{n-2}t\right),$$

$$(s=1,\ldots,n-2)$$

$$b^{(n-1)} = \left(\binom{n}{1}t,\ \binom{n}{2}Dt,\ldots,\binom{n}{n-1}D^{n-2}t\right),$$

$$t = d^{-1}D$$

for $d > 1$ and

$$b^{(0)} = \left(\binom{n}{1}D,\ \binom{n}{2}D^2, \ldots, \binom{n}{n-1}D^{n-1} \right)$$

for $d = 1$. The JPA is ideally convergent for $D \geq \frac{3}{2}(n-1)d$.

The reader should note that with $d|D$, $\sqrt[n]{D^n + d}$ is an n-th degree irrational, as can be easily proved. The irreducibility criterion for $x^n - (D^n + d) = 0$, here given by $D \geq \frac{3}{2}(n-1)d$, is only a sufficient condition, as was already outlined before. The n-th degree irrational $\alpha = \sqrt[n]{D^n + d}$ can now be rationally approximated by formula (3.53). We obtain

$$w = \alpha - D = \lim_{s \to \infty} \frac{A_0^{(sn)}}{A_0^{(sn+1)}},$$

(4.12)
$$\alpha = D + \lim_{s \to \infty} \frac{A_0^{(sn)}}{A_0^{(sn+1)}}.$$

An explicit formula for $A_0^{(v)}$, expressed as a function of D and d, will be given in a next chapter.

Example 7. $\alpha = \sqrt[4]{D^4 + d};\ d|D.$

$$Q(\alpha) \ni a^{(0)} = (w+3D, w^2+2Dw+3D^2, w^3+Dw^2+D^2w+D^3)$$

$$b^{(0)} = (4D,\ 6D^2,\ 4D^3),$$

$$b^{(1)} = (4D,\ 6D^2,\ 4D^2t),$$

$$b^{(2)} = (4D,\ 6Dt,\ 4D^2t),$$

$$b^{(3)} = (4t,\ 6Dt,\ 4D^2t).$$

$A_0^{(4)} = 1;\ A_0^{(5)} = 4D^2t;\ A_0^{(6)} = 16D^4t^2+6Dt;\ A_0^{(7)} = 64D^6t^3+48D^3t^2+4t;$

$A_0^{(8)} = 256D^9t^3+288D^6t^2+68D^3t+1;\ A_0^{(9)} = 1024D^{11}t^4+1536D^8t^3+624D^5t^2+56D^2t.$

$$\sqrt[4]{D^4+d} \approx \frac{256D^9t^3+288D^6t^2+68D^3t + 1}{D^2t(1024D^9t^3+1536D^6t^2+624D^3t+56)} + D \approx D + \frac{1}{4D^2t}.$$

Most of the classical investigations of periodicity of the JPA with $f(a^{(k)}) = \left[a^{(k)}\right]$ concentrated on a vector $a^{(0)} \in Q(\alpha)$ of the form

$$a^{(0)} = (\alpha, \alpha^2, \ldots, \alpha^{n-1}) \in E_{n-1},$$

which may seem a natural approach with the numbers $1, \alpha, \alpha^2, \ldots, \alpha^{n-1}$ forming a basis of $Q(\alpha)$. As we shall see, the JPA of such $a^{(0)}$ also becomes periodic, but has a pre-period of length n-1. But for both theoretical and practical ends, it is not necessary to drag along, so to say, the pre-period; if the length of the pre-period is ℓ, one can start, instead of using $a^{(0)}$, with the vector $a^{(\ell)}$ and obtain a purely periodic JPA. This was done in Theorem 4. We leave to the reader the proof of the completely analogous

Corollary 1 (to Theorem 4). Let α have the structure as in Theorem 4, and let

$$a^{(0)} = (\alpha, \alpha^2, \ldots, \alpha^{n-1}), \tag{4.13}$$

then the JPA of $a^{(0)}$ with $f(a^{(k)}) = [a^{(k)}]$ is periodic; the length of the pre-period $\ell = n-1$; the length of the period is n for $d > 1$ and 1 for $d = 1$. The pre-period has the form

$$(4.14)\quad \begin{cases} b^{(0)} = (D, D^2, \ldots, D^{n-1}) \\ b^{(s)} = \left(\binom{s+1}{s} D, \binom{s+2}{s} D^2, \ldots, \binom{n-1}{s} D^{n-s-1}, \binom{n}{n-s}\frac{D^{n-s}}{d}, \right. \\ \qquad\qquad \left. \binom{n}{n-s+1}\frac{D^{n-s+1}}{d}, \ldots, \binom{n}{n-1}\frac{D^{n-1}}{d} \right) \\ \qquad\qquad (s=1,\ldots,n-2) \end{cases}$$

For $d > 1$, the period has the form

$$(4.15)\quad \begin{cases} b^{(n-1)} = \left(\binom{n}{1} t, \binom{n}{2} Dt, \ldots, \binom{n}{n-1} D^{n-2} t \right), \\ b^{(n)} = \left(\binom{n}{1} D, \binom{n}{2} D^2, \ldots, \binom{n}{n-1} D^{n-1} \right), \\ b^{(n+k)} = \left(\binom{n}{1} D, \binom{n}{2} D^2, \ldots, \binom{n}{n-k-1} D^{n-k-1}, \binom{n}{n-k} D^{n-k-1} t, \ldots, \right. \\ \qquad\qquad \left. \binom{n}{n-1} D^{n-2} t \right). \\ \qquad\qquad (k=1,\ldots,n-2;\ t=d^{-1}D) \end{cases}$$

For $d = 1$, the period has the form

$$(4.16) \qquad b^{(n-1)} = \left(\binom{n}{1}D, \binom{n}{2}D^2, \dots, \binom{n}{n-1}D^{n-1} \right).$$

It is now obvious that if we add the vector $b^{(n-1)}$ to the pre-period, and take for the period the vectors $b^{(n)}$, $b^{(n+1)}$, $b^{(2n-1)}$, then the period has the same form as in Theorem 4.

If in (4.14), (4.15) we formally substitute $n = p^v$ (p a prime, $v \geq 1$) and $d = pd'$, $d' | D$, then the components of these vectors are again integers, since

$$p \,\Big|\, \binom{p^v}{s}, \qquad s = 1, \dots, p^v - 1, \ v \geq 1.$$

That with these substitutions, the JPA of $a^{(0)}$ with $f(a^{(k)}) = [a^{(k)}]$ becomes indeed periodic with the transformed pre-period and period, can be verified by the reader without any special effort, and is expressed by

Corollary 2 (to Theorem 4). Let

$$\alpha' = \sqrt[n]{D^n + pd}, \ n = p^v, \ p \text{ a prime}, \ v \geq 1; \ d | D.$$

$$D \geq (e-2)(n-1)pd, \ n \leq 4. \quad D \geq (n-2)pd, \ n > 4.$$

The JPA with $f(a^{(k)}) = [a^{(k)}]$ of

$$a^{(0)} = (\alpha', \alpha'^2, \dots, \alpha'^{n-1}) \in Q(\alpha')$$

is periodic; its pre-period and period have the same form as in (4.14) and (4.15), where n is to be substituted by p^v, and d by pd. The length of the period is always n.

§ 2. Periodicity of $a^{(0)} \in E_{n-1}$, $\alpha = (D^n - \frac{d}{m})^{\frac{1}{n}}$

In this paragraph we shall investigate periodicity of the JPA with $f(a^{(k)}) = [a^{(k)}]$ of a vector selected with components in an algebraic field generated by the irrational $\alpha = (D^n - \frac{d}{m})^{\frac{1}{n}}$ over the rationals. For $m = 1$, α becomes $\sqrt[n]{D^n - d}$ which is, so to say, the counterpart of $\alpha = \sqrt[n]{D^n + d}$.

We shall use the following notations: an n x (n-1) matrix of the form

$$(4.17)\qquad \begin{pmatrix} A_1, & A_2, & \dots, & A_{n-1} \\ 0 & & & 1 \\ & \ddots & & \vdots \\ & & 0 & 1 \end{pmatrix}$$

will be called a fugue. The first row of a fugue will be called its accumulator, and the numbers $A_1,\dots,A_{n-1}$ – the elements of the accumulator. We shall use the generalized sigma sign

$$(4.18)\qquad \sum_{i=0|c}^{t-1|n} a_i \equiv c\sum_{i=0}^{t-1} a_i + \sum_{i=t}^{n} a_i.$$

The main result of this paragraph is stated in

THEOREM 5. Let α be the irrational

$$(4.19)\qquad \alpha = \left(D^n - \frac{d}{m}\right)^{\frac{1}{n}};\ m,n,d,D \in N;\ d|D;\ D \geqslant 2(n-1)d.$$

Then the JPA, with $f(a^{(k)}) = \left[a^{(k)}\right]$, of the vector

$$(4.20)\qquad a^{(0)} = \left(\dots, \sum_{i=0}^{s} \binom{n-1-s+i}{i}(D-1)^i w^{s-i}, \dots\right) \qquad (s=1,\dots,n-1)$$

is purely periodic and its primitive length equals n^2, if not both m and d = 1, and n if d = m = 1. The period consists (for m,d not both 1) of n fugues. The accumulator of the first fugue has the form

$$(4.21)\qquad A_k = -1 + \sum_{i=0}^{k} \binom{n-1-k+i}{i}(D-1)^i D^{k-i}. \qquad (k=1,\dots,n-1)$$

The accumulator of the s-th fugue (s = 2,...,n-1) has the form: the first n-s elements of the accumulator are given by

$$(4.22)\qquad A_k = -1 + \sum_{i=0}^{k} \binom{n-1-k+i}{i}(D-1)^i D^{k-i}; \qquad (k=1,\dots,n-s)$$

and the succeeding s-1 elements by

$$(4.23) \qquad A_{n-s+t} = -1 + \sum_{i=0 \mid d^{-1}m}^{t-1 \mid n-s+t} (-1)^i \binom{s-1-t+i}{i} \binom{n}{s-t+i} D^{n-s+t-i}.$$

$$(t=1,\ldots,s-1)$$

The n-1 elements of the accumulator of the n-th fugue have the form

$$(4.24) \qquad A_t = -1 + \sum_{i=0 \mid d^{-1}m}^{t-1 \mid t} (-1)^i \binom{n-1-t+i}{i} \binom{n}{t-i} D^{t-i}. \quad (t=1,\ldots,n-1)$$

In the case $m = d = 1$, the primitive period consists of one fugue, and its accumulator has the form (4.21).

This rather complicated Theorem was proved by the author in [2,c].

We shall illustrate Theorem 5 for the cases $n = 2$ and $n = 5$.

(i) $n = 2;\ \alpha = (D^2 - d^{-1}m)^{\frac{1}{2}};\ d \mid D;\ D \geqslant 2d.$

$$a^{(0)} = \alpha + (D-1).$$

The fugues have here the form

$$\begin{pmatrix} A_1 \\ 1 \end{pmatrix}.$$

The element of the first fugue, according to (4.21) has the form

$$A_1 = -1 + \sum_{i=0}^{1} (D-1)^i D^{1-i} = 2(D-1);$$

the element of the second (last) fugue has the form, according to (4.24)

$$A_1 = -1 + \sum_{i=0 \mid d^{-1}m}^{0 \mid 1} (-1)^i \binom{2}{1-i} D^{1-i} = 2(md^{-1}D-1);$$

therefore

$$(4.25) \qquad \sqrt{D^2 - d^{-1}m} + (D-1) = \left[\overline{2(D-1),\ 1,\ 2(md^{-1}D-1),\ 1}\right].$$

The reader can verify (4.25) easily by means of simple continued fractions.

(ii) $n = 5;\ \alpha = (D^5 - d^{-1}m)^{\frac{1}{5}};\ d|D;\ D \geqslant 8d.$

The components of $a^{(0)}$ are

$$a_1^{(0)} = \alpha+4(D-1);\ a_2^{(0)} = \alpha^2+3(D-1)\alpha + 6(D-1)^2;$$

$$a_3^{(0)} = \alpha^3 + 2(D-1)\alpha^2 + 3(D-1)^2\alpha + 4(D-1)^3;$$

$$a_4^{(0)} = \alpha^4 + (D-1)\alpha^3 + (D-1)^2\alpha^2 + (D-1)^3\alpha + (D-1)^4.$$

The reader will verify easily, on the basis of (4.21)-(4.24) that the accumulators of the five fugues have the form

First fugue:

$5(D-1);\ 5(D-1)(2D-1);\ 5(D-1)(2D^2-2D+1);\ 5D(D-1)(D^2-D+1);$

Second fugue:

$5(D-1);\ 5(D-1)(2D-1);\ 5(D-1)(2D^2-2D+1);\ 5D(d^{-1}mD^3-2D^2+2D-1);$

Third fugue:

$5(D-1);5(D-1)(2D-1);5(2d^{-1}mD^3-4D^2+3D-1);5D\left[d^{-1}mD^2(D-2)+2D-1\right];$

Fourth fugue:

$5(D-1);5(2d^{-1}mD^2-3D+1);5\left[2d^{-1}mD^2(D-2)+3D-1\right];5D\left[d^{-1}m(D^3-2D^2+2D)-1\right];$

Fifth fugue:

$5(d^{-1}mD-1);5\left[d^{-1}m(2D^2-3D)+1\right];\ 5\left[d^{-1}m(2D^2-4D^2+3D)-1\right];$

$$5d^{-1}m(D^4-2D^3+2D^2-D).$$

Before dealing with the proof of Theorem 5, we shall introduce a few notations, identities and equalities which are the main tools for our treatment of the theorem. We denote

$$(4.26)\qquad a_s^{(0)} \equiv f_s(\alpha,D-1) = \sum_{i=0}^{s}\binom{n-s-1+i}{i}\alpha^{s-i}(D-1)^i.$$

$$(s=1,\dots,n-1;\ f_0(\alpha,D-1)=1)$$

$$(4.27)\qquad F_s(\alpha,D) = \sum_{i=0}^{s}\binom{n-s-1+i}{i}\alpha^{s-i}D^i.\qquad (s=1,\dots,n-1;F_0(\alpha,D)=1)$$

$$(4.28)\qquad g_{n-s,t}(\alpha,D) = \sum_{i=0\,|\,d^{-1}m}^{n-1\,|\,n-s+t} (-1)^i \binom{s-1-t+i}{i} F_{n-s+t-i}(\alpha,D).$$

$$(s=2,3,\ldots,n;\ t=1,2,\ldots,s-1)$$

For any polynomial in α,D with integral coefficients, viz

$$(4.29)\qquad P_s(\alpha,D) = \sum_{i=0}^{s} c_i\alpha^{s-i}D^i \qquad (c_i \in I;\ s=1,\ldots,n-1; P_0=1)$$

the following abbreviations will be used:

$$(4.30)\qquad P_s(\alpha,D)=P_s;\quad P_s(D,D) = \bar{P}_s;\quad \frac{\bar{P}_s - P_s}{\bar{P}_1 - P_1} = {}^{(1)}P_s.$$
$$(s=1,\ldots,n-1;\ {}^{(1)}P_0=0)$$

We shall enumerate a series of identities and leave their verification to the reader; the proof amounts mostly to summation of binomial coefficients.

$$(4.31)\qquad f_s(D-1,D-1) = \binom{n}{s}(D-1)^s. \qquad (s=0,1,\ldots,n-1)$$

$$(4.32)\qquad \bar{F}_s = \binom{n}{s}D^s. \qquad (s=0,1,\ldots,n-1)$$

$$(4.33)\qquad {}^{(1)}F_s = F_{s-1}. \qquad (s=1,\ldots,n-1)$$

$$(4.34)\qquad f_s = \sum_{i=0}^{s}(-1)^i\binom{n-1-s+i}{i}F_{s-1}. \qquad (s=0,1,\ldots,n-1)$$

$$(4.35)\qquad {}^{(1)}f_s = \sum_{i=0}^{s-1}(-1)^i\binom{n-1-s+i}{i}F_{s-1-i}. \qquad (s=1,\ldots,n-1)$$

$$(4.36)\qquad {}^{(1)}f_s - {}^{(1)}f_{s-1} = f_{s-1}. \qquad (s=1,\ldots,n-1)$$

$$(4.37)\qquad {}^{(1)}g_{n-s,t} = \sum_{i=0\,|\,m^{-1}d}^{t-1\,|\,n-s+t-1} (-1)^i\binom{s-t-1+i}{i}F_{n-s+t-1-i}.$$

$$(s=2,\ldots,n;\ t=1,\ldots,s-1)$$

where, as before, we mean by ${}^{(1)}g_{n-s,t}$ the expression

$$\frac{\bar{g}_{n-s,t} - g_{n-s,t}}{D - \alpha}.$$

(4.38) $\quad {}^{(1)}g_{n-s,1} - {}^{(1)}f_{n-s} = g_{n-s-1,1}. \qquad (s=1,\ldots,n-1)$

(4.39) $\quad {}^{(1)}g_{n-s,t+1} - {}^{(1)}g_{n-s,t} = g_{n-s-1,t+1}.$

$(s=2,\ldots,n-1;\ t=1,\ldots,s-1)$

From (4.39) we obtain, of course,

(4.40) $\quad {}^{(1)}g_{n-s,t} - {}^{(1)}g_{n-s,q} = \sum_{i=0}^{t-1-q} g_{n-s-1,t-i}. \qquad (q < t)$

(4.41) $\quad {}^{(1)}g_{0,t} - {}^{(1)}g_{0,t-1} = d^{-1}mf_{t-1}. \qquad (t=1,\ldots,n-1)$

(4.42) $\quad \dfrac{g_{n-s,s-1} - \bar{g}_{n-s,s-1} + 1}{D - \alpha} = g_{n-s-1,s}. \qquad (s=2,\ldots,n-1)$

(4.43) $\quad \dfrac{g_{0,n-1} - \bar{g}_{0,n-1} + 1}{d^{-1}m(D - \alpha)} = f_{n-1}.$

(4.44) $\quad {}^{(1)}f_s - {}^{(1)}f_q = \sum_{i=0}^{s-q-1} f_{s-1-i}. \qquad (1 \leqslant q \leqslant s-1)$

(4.45) $\quad {}^{(1)}g_{n-s,t} - {}^{(1)}g_q = \sum_{i=0}^{t-1} g_{n-(s+1),t-i} + \sum_{i=0}^{n-s-q-1} f_{n-s-1-i}.$

$(1 \leqslant q \leqslant n-s)$

We shall now prove a series of important inequalities.

(4.46) $\quad (D-1)^k < \alpha^k < D^k. \qquad (k=1,\ldots,n-1)$

This follows immediately from the very structure of α. From (4.46) and the definition of f_s and F_s it further follows

(4.47) $\quad f_s(D-1,D-1) < f_s < F_s < \bar{F}_s;\ f_s < \bar{f}_s < \bar{F}_s. \quad (s=1,\ldots,n-1)$

(4.48) $\quad (1 + \frac{1}{D-1})^{n-2} < 1.65.$

Since $D \geqslant 2d(n-1) \geqslant 2(n-1)$, we obtain

$$(1+\frac{1}{D+1})^{n-2} < (1+\frac{1}{2(n-2)})^{n-2} = \left[(1+\frac{1}{2(n-2)})^{2(n-2)}\right]^{\frac{1}{2}} < e^{\frac{1}{2}} < 1.65.$$

We use similar techniques to prove the following inequalities which the reader will verify easily

(4.49) $\bar{F}_i \leqslant F_{i+1}(D-1,D-1).$ $(i=0,1,\ldots,n-2)$

(4.50) $F_s < F_{s+t};\ f_s < f_{s+t}.$ $(s=0,\ldots,n-2; t=1,\ldots,n-1;$
$1 \leqslant s+t \leqslant n-1)$

(4.51) $2F_{n-2} < d^{-1}F_{n-1}.$

(4.52) ${}^{(1)}f_s \leqslant F_{s-1}.$ $(s=1,\ldots,n-1)$

(4.53) ${}^{(1)}g_{n-s,t} < d^{-1}mF_{n-s+t-1}.$ $(s=2,\ldots,n;\ t=1,\ldots,s-1)$

(4.54) $g_{n-s,t} < d^{-1}mF_{n-s+t}.$ $(s=2,\ldots,n;\ t=1,\ldots,s-1)$

(4.55) $[f_s] = -1 + \bar{f}_s.$ $(s=1,\ldots,n-1)$

To prove (4.55) we must show

a) $-1 + \bar{f}_s < f_s$; b) $f_s < \bar{f}_s$.

a) We must prove: $\bar{f}_s - f_s < 1$, or, dividing by $D - \alpha > 0$,

$${}^{(1)}f_s = \frac{1}{D-\alpha} = \frac{D^{n-1} + D^{n-2}\alpha + \cdots + \alpha^{n-1}}{D^n - \alpha^n}$$

$$= \frac{F_{n-1}}{D^n - (D^n - dm^{-1})} = d^{-1}mF_{n-1}.$$

But from (4.51), (4.52) we obtain

$${}^{(1)}f_s \leqslant F_{s-1} \leqslant F_{n-2} < d^{-1}mF_{n-1}.$$

b) This follows from (4.47). In a similar way we prove:

(4.56) $\quad [g_{n-s,t}] = -1 + \bar{g}_{n-s,t}.$ $\qquad$ $(s=2,\ldots,n;t=1,\ldots,s-1)$

(4.57) $\quad 0 < \dfrac{f_i - \bar{f}_i + 1}{f_j - \bar{f}_j + 1} < 1.$ $\qquad$ $(j \leqslant i \leqslant n-1;\ j=1,\ldots,n-2)$

(4.58) $\quad 0 < \dfrac{(D-\alpha)f_j}{f_s - \bar{f}_s + 1} < 1.$ $\qquad$ $(j=0,\ldots,s-2;s=2,\ldots,n-1)$

(4.59) $\quad 1 < \dfrac{f_{i-1} - \bar{f}_{k-1} + 1}{f_i - \bar{f}_i + 1} < 2.$ $\qquad$ $(i=1,\ldots,n-1)$

(4.60) $\quad 0 < \dfrac{g_{n-s,t} - \bar{g}_{n-s,t} + 1}{f_q - \bar{f}_q + 1} < 1.$

$(s=2,\ldots,n;\ t=1,\ldots,s-1;\ q=1,\ldots,n-s)$

(4.61) $\quad 0 < \dfrac{g_{n-s,t} - \bar{g}_{n-s,t} + 1}{g_{n-s,j} - \bar{g}_{n-s,j} + 1} < 1.$ $\qquad$ $(1 \leqslant j \leqslant t \leqslant s-1 \leqslant n-1)$

(4.62) $\quad 1 < \dfrac{f_{n-s} - \bar{f}_{n-s} + 1}{g_{n-s,1} - \bar{g}_{n-s,1} + 1} < 2.$ $\qquad$ $(s=2,\ldots,n-1)$

(4.63) $\quad 0 < \dfrac{(D-\alpha)f_{r-1}}{g_{n-s,t} - \bar{g}_{n-s,t} + 1} < 1.$

$(r=1,\ldots,n-s;\ s=2,\ldots,n-1;\ t=1,\ldots,s-1)$

(4.64) $\quad 1 < \dfrac{g_{n-s,t} - \bar{g}_{n-s,t} + 1}{g_{n-s,t+1} - \bar{g}_{n-s,t+1} + 1} < 2.$ $\qquad$ $(s=2,\ldots,n;t=1,\ldots,s-2)$

We are now sufficiently equipped with identities and inequalities to prove Theorem 5. We shall successively construct $a^{(0)}T^k$ $(k=1,\ldots,n)$ and derive $f(a^{(k)}) = \left[a^{(k)}\right] = b^{(k)}$, thus obtaining the first fugue and the accumulator of the second fugue. On basis of the methods developed in these operations, the reader will have no difficulty to verify the structure of the remaining n-1 fugues.

Having introduced the notation $a_i^{(0)} = f_i$, we obtain, in virtue of (4.55),

$$(4.65) \qquad b_k^{(0)} = [f_k] = -1 + \bar{f}_k$$

and recalling the meaning of $\bar{f}_i$ from (4.30)

$$(4.66) \qquad b_k^{(0)} = -1 + \sum_{i=0}^{k} \binom{n-1-k+i}{i} (D-1)^i D^{k-i}. \qquad (k=1,\dots,n-1)$$

Comparing (4.66) with (4.22), we see that this vector $b^{(0)}$ is really the accumulator of the first fugue. We further obtain, from (4.26) and (4.66),

$$(4.67) \qquad \begin{aligned} a^{(1)} &= \frac{1}{f_1-\bar{f}_1+1} (f_2-\bar{f}_2+1,\dots,f_{n-1}-\bar{f}_{n-1}+1,\ 1), \\ a_{i-1}^{(1)} &= \frac{f_i-\bar{f}_i+1}{f_1-\bar{f}_1+1} \ (i=2,\dots,n-1);\ a_{n-1}^{(1)} = \frac{1}{f_1-\bar{f}_1+1}, \end{aligned}$$

and from (4.67), in virtue of (4.57) and (4.59)

$$(4.68) \qquad b_i^{(1)} = 0, \ (i=1,\dots,n-2) \qquad b_{n-1}^{(1)} = 1$$

which supplies the second row-vector of the first fugue.

From (4.67), (4.68) we obtain

$$a^{(2)} = \frac{f_1-\bar{f}_1+1}{f_2-\bar{f}_2+1} \left(\frac{f_3-\bar{f}_3+1}{f_1-\bar{f}_1+1},\dots,\frac{f_{n-1}-\bar{f}_{n-1}+1}{f_1-\bar{f}_1+1},\ \frac{\bar{f}_1-f_1}{f_1-\bar{f}_1+1},\ 1 \right)$$

so that

$$a^{(2)} = \frac{f_i - \bar{f}_i + 1}{f_2 - \bar{f}_2 + 1}, \qquad (i=3,\dots,n-1)$$

$$a_{n-2}^{(2)} = \frac{\bar{f}_1 - f_1}{f_2 - \bar{f}_2 + 1}, \qquad a_{n-1}^{(2)} = \frac{f_1 - \bar{f}_1 + 1}{f_2 - \bar{f}_2 + 1}.$$

Since $f_1 = \alpha + (n-1)(D-1)$, $\bar{f}_1 = D + (n-1)(D-1)$, we have

$$\bar{f}_1 - f_1 = D - \alpha = (D-\alpha)f_0,$$

so that the above formulas take the form

$$(4.69)\qquad \begin{aligned} a_{i-2}^{(2)} &= \frac{f_i-\bar{f}_i+1}{f_2-\bar{f}_2+1}, \qquad (i=3,\dots,n-1) \\ a_{n-2}^{(2)} &= \frac{(D-\alpha)f_0}{f_2-\bar{f}_2+1}, \qquad a_{n-1}^{(2)} = \frac{f_1-\bar{f}_1+1}{f_2-\bar{f}_2+1}. \end{aligned}$$

From (4.69) we obtain, in virtue of (4.57), (4.58) and (4.59)

$$(4.70)\qquad b_i^{(2)} = 0 \quad (i=1,\dots,n-2); \qquad b_{n-1}^{(2)} = 1.$$

From (4.69), (4.70) we now obtain, as before,

$$a_{i-3}^{(3)} = \frac{f_i - \bar{f}_i + 1}{f_3 - \bar{f}_3 + 1}, \qquad (i=4,\dots,n-1)$$

$$a_{n-3}^{(3)} = \frac{(D-\alpha)f_0}{f_3-\bar{f}_3+1}, \quad a_{n-2}^{(3)} = \frac{(\bar{f}_2-f_2)-(\bar{f}_1-f_1)}{f_3 - \bar{f}_3 + 1}, \quad a_{n-1}^{(3)} = \frac{f_2-\bar{f}_2+1}{f_3-\bar{f}_3+1}.$$

Now, $\bar{f}_2-f_2 = (D-\alpha)^{(1)}f_2$, $\bar{f}_1-f_1 = D-\alpha = (D-\alpha)^{(1)}f_1$, $(\bar{f}_2-f_2)-(\bar{f}_1-f_1) = (D-\alpha)({}^{(1)}f_2-{}^{(1)}f_1) = (D-\alpha)f_1$, by formula (4.36). We have thus obtained

$$(4.71)\qquad \begin{aligned} a_{i-3}^{(3)} &= \frac{f_i - \bar{f}_i + 1}{f_3 - \bar{f}_3 + 1}, \qquad (i=4,\dots,n-1) \\ a_{n-3}^{(3)} &= \frac{(D-\alpha)f_0}{f_3-\bar{f}_3+1}, \quad a_{n-2}^{(3)} = \frac{(D-\alpha)f_1}{f_3-\bar{f}_3+1}, \quad a_{n-1}^{(3)} = \frac{f_2-\bar{f}_2+1}{f_3-\bar{f}_3+1}. \end{aligned}$$

There will be no difficulty now to prove the following formula

$$
(4.72)\qquad \begin{cases} a_{i-k}^{(k)} = \dfrac{f_i - \bar{f}_i + 1}{f_k - \bar{f}_k + 1}, & (i=k+1,\dots,n-1) \\[2ex] a_{n-k+j}^{(k)} = \dfrac{(D-\alpha)f_j}{f_k - \bar{f}_k + 1}, & (j=0,\dots,k-2) \\[2ex] a_{n-1}^{(k)} = \dfrac{f_{k-1} - \bar{f}_{k-1} + 1}{f_k - \bar{f}_k + 1}, & (k=2,\dots,n-2) \end{cases}
$$

Formula (4.72) is correct for $k=2,3$, as can be easily verified because of (4.69) and (4.71). Proof then follows by induction.

We also obtain from (4.72), in virtue of (4.57), (4.58), (4.59)

$$
(4.73)\qquad b_i^{(k)} = 0 \ (i=1,\dots,n-2); \quad b_{n-1}^{(k)} = 1 \quad (k=2,\dots,n-2).
$$

From (4.72) we obtain, for $k = n-2$,

$$
(4.74)\qquad \begin{aligned} & a_1^{(n-2)} = \frac{f_{n-1}-\bar{f}_{n-1}+1}{f_{n-2}-\bar{f}_{n-2}+1}, \qquad a_{2+j}^{(n-2)} = \frac{(D-\alpha)f_j}{f_{n-2}-\bar{f}_{n-2}+1}, \quad (j=0,1,\dots,n-4) \\ & a_{n-1}^{(n-2)} = \frac{f_{n-3}-\bar{f}_{n-3}+1}{f_{n-2}-\bar{f}_{n-2}+1}, \end{aligned}
$$

and from (4.74), in virtue of (4.57), (4.58), (4.59)

$$
(4.75)\qquad b_i^{(n-2)} = 0 \quad (i=1,\dots,n-2); \quad b_{n-1}^{(n-2)} = 1.
$$

From (4.74), (4.75) we obtain, as before,

$$
(4.76)\qquad \begin{aligned} & a_{1+j}^{(n-1)} = \frac{(D-\alpha)f_j}{f_{n-1}-\bar{f}_{n-1}+1} \qquad (j=0,1,\dots,n-3) \\ & a_{n-1}^{(n-1)} = \frac{f_{n-2}-\bar{f}_{n-2}+1}{f_{n-1}-\bar{f}_{n-1}+1}, \end{aligned}
$$

and from (4.76), in virtue of (4.58), (4.59)

$$
(4.76a)\qquad b_i^{(n-1)} = 0 \quad (i=1,\dots,n-2); \quad b_{n-1}^{(n-1)} = 1
$$

so that, since $f_0 = 1$,

$$a^{(n)} = \frac{f_{n-1}-\bar{f}_{n-1}+1}{D-\alpha}\left(\frac{(D-\alpha)f_1}{f_{n-1}-\bar{f}_{n-1}+1}, \ldots, \frac{(D-\alpha)f_{n-2}}{f_{n-1}-\bar{f}_{n-1}+1}, 1\right)$$

(4.77)
$$a_i^{(n)} = f_i \qquad (i=1,\ldots,n-2)$$
$$a_{n-1}^{(n)} = \frac{f_{n-1}-\bar{f}_{n-1}+1}{D-\alpha}.$$

We shall take a better look at $a_{n-1}^{(n)}$ and obtain

$$a_{n-1}^{(n)} = \frac{1}{D-\alpha} - \frac{\bar{f}_{n-1}-f_{n-1}}{D-\alpha} = \frac{D^{n-1}+D^{n-2}\alpha+\cdots+\alpha^{n-1}}{D^n - \alpha^n} - {}^{(1)}f_{n-1}$$

$$= d^{-1}mF_{n-1} - \sum_{i=0}^{n-2}(-1)^i\binom{i}{i}F_{n-2-i},$$

following formula (4.35), where $s=n-1$. Thus

$$a_{n-1}^{(n)} = \sum_{i=0|d^{-1}m}^{(n-1)}(-1)^i F_{n-1-i} = g_{n-2,1},$$

(4.78)
$$a_i^{(n)} = f_i \quad (i=1,\ldots,n-2); \quad a_{n-1}^{(n)} = g_{n-2,1}.$$

Now, from (4.56): $[g_{n-2,1}] = -1 + \bar{g}_{n-2,1}$, from (4.28): $\bar{g}_{n-2,1} = \sum_{i=0|d^{-1}m}^{0|n-1}(-1)^i\bar{F}_{n-1-i}$, and from (4.32): $\bar{F}_{n-1-i} = \binom{n}{n-1-i}D^{n-1-i}$, so that

$$\left[a_{n-1}^{(n)}\right] = -1 + \sum_{i=0|d^{-1}m}^{0|n-1}(-1)^i\binom{n}{1+i}D^{n-1-i},$$

and since $[f_i] = -1 + \bar{f}_i$ as before, we obtain

(4.79)
$$b_k^{(n)} = -1 + \sum_{i=0}^{k}\binom{n-1-k+i}{i}(D-1)^i D^{k-i}; \qquad (k=1,\ldots,n-2)$$
$$b_{n-1}^{(n)} = -1 + \sum_{i=0|d^{-1}m}^{0|n-1}(-1)^i\binom{n}{1+i}D^{n-1-i}.$$

Thus the row vector $b^{(n)}$ is indeed the accumulator of the second fugue. We have verified, in virtue of (4.66), (4.68), (4.73), (4.76a) and (4.79), that the vectors $b^{(k)}$ $(k=0,\ldots,n)$ form indeed the first fugue and the accumulator of the second fugue of the period, as demanded by Theorem 5. The theorem is then easily proved by induction.

The surprising length of the period of the JPA $a^{(0)} = a^{(0)}(\alpha)$, $\alpha = \sqrt[n]{D^n - d}$, namely n^2, **does not yet justify the conjecture** that the JPA of any $a^{(0)}$ with $f(a^{(k)}) = \left[a^{(k)}\right]$ may always become periodic. This was, indeed, the hope that had animated first C. G. J. Jacobi and later O. Perron. Recent calculations for $n = 3,4,5$ that have been computerized by the author and published in [2,d] seem to support this conjecture. The author has also quite recently developed entirely new methods by means of which he hopes to prove general periodicity, or, at least, to show when periodicity would occur and when not. If $\alpha = \sqrt[n]{D^n - d}$, and one starts with the vector $a^{(0)} = (\alpha,\alpha^2,\ldots,\alpha^{n-1})$, as in the case of $\alpha = \sqrt[n]{D^n + d}$, then the JPA of $a^{(0)}$ with $f(a^{(k)}) = \left[a^{(k)}\right]$ becomes again periodic, and the length of the pre-period is $\ell = (n-1)^2 + 1$, as was proved by the author in [2,g].

The reader should note that Theorem 5 also covers the case

$$\alpha = \sqrt[n]{D^n - pd},\quad n = p^v \quad (v > 1;\ p \text{ a prime}).$$

§ 3. The cases $\alpha = \sqrt[3]{D^3 + 3D}$ and $\alpha = \sqrt[3]{D^3 + 6D}$

According to Corollary 2 of paragraph 1 of this chapter, the JPA with $f(a^{(k)}) = \left[a^{(k)}\right]$ of the vector $a^{(0)} = (\alpha, \alpha^2)$, $\alpha = \sqrt[3]{D^3 + 3d}$, $d|D$, becomes periodic, if $D \geqslant 3(e-2)(n-1)d$. This condition does not hold, of course, for $d = D$. It is therefore the more surprising that the JPA with $f(a^{(k)}) = \left[a^{(k)}\right]$ of $a^{(0)} = (\alpha,\alpha^2)$, $\alpha = \sqrt[3]{D^3 + 3D}$ still becomes periodic, though, as could be expected, the structure of the period and the pre-period are entirely different from those of Corollary 2. This result was obtained by the author in [2,e] and is stated in

THEOREM 6. Let $D \geqslant 2$ be a natural number and let denote

$$\alpha = \sqrt[3]{D^3 + 3D}. \tag{4.80}$$

Then the JPA with $f(a^{(k)}) = \left[a^{(k)}\right]$ of

$$a^{(0)} = (\alpha, \alpha^2) \tag{4.81}$$

becomes periodic; the length of the primitive pre-period is four, and the pre-period has the form

$$b^{(0)}=(D,D^2+1);\ b^{(1)}=(D-1,D);\ b^{(2)}=(0,1);\ b^{(3)}=(0,D); \tag{4.82}$$

the length of the primitive period is also four, and the period has the form

$$b^{(4)}=(2D-1,3D^2+3);\ b^{(5)}=(D-1,D);\ b^{(6)}=(0,1);\ b^{(7)}=(1,D-1). \tag{4.83}$$

The condition of Theorem 6, viz. $D \geqslant 2$, is most essential; for $D = 1$ the theorem is not valid. In this case $\alpha = \sqrt[3]{4}$, $a^{(0)} = (\sqrt[3]{4}, \sqrt[3]{16})$. This vector seems to occupy a most magic, and most annoying, place in the theory of JPA with $f(a^{(k)}) = \left[a^{(k)}\right]$. No human effort nor the capability of any computer have ever succeeded to find even a small hint of periodicity, tests having been pushed up to $a^{(0)}T^{150}$. Of course, this does not yet disprove its periodicity.

The proof of Theorem 6 consists in carrying out the successive transformations $a^{(0)}T^k$, each one consisting of the following three steps:

(i) starting with $a^{(k)} = (a_1^{(k)}, a_2^{(k)})$; $(k=0,1,\ldots)$

(ii) calculating $b^{(k)} = \left(\left[a_1^{(k)}\right], \left[a_2^{(k)}\right]\right)$;

(iii) rationalizing the denominator of the components of

$$a^{(k+1)} = \frac{1}{a_1^{(k)}-b_1^{(k)}}\left(a_2^{(k)}-b_2^{(k)},\ 1\right).$$

A machinery of formulas can be easily hammered out for this purpose; this was done by the author in [2,f]. With these, the reader will find no difficulty to prove Theorem 6. A difficult operation is the evaluation of $\left[a^{(k)}\right]$; here one should be guided by the inequalities

$$\frac{1}{D} - \frac{1}{D^3} < \alpha - D = \sqrt[3]{D^3 + 3D} - D < \frac{1}{D} - \frac{1}{D^3} + \frac{5}{3} \cdot \frac{1}{D^5}.$$

For a better orientation of the reader in proving Theorem 6, we shall state here only the eight vectors $a^{(k)}$, $(k=0,1,\ldots,7)$ from which the pre-period (4.82) and the period (4.83) are derived.

$$a^{(0)} = (\alpha,\ \alpha^2), \qquad\qquad (\alpha = \sqrt[3]{D^3 + 3D}\,)$$

$$a^{(1)} = \frac{1}{3D}(-\alpha^2+2D\alpha+2D^2,\ \alpha^2+D\alpha+D^2),$$

$$a^{(2)} = \left(\frac{-2(D-1)\alpha^2+(D+1)^2\alpha+(D-1)(D^2-1)}{3D^3-3D^2+9D-1},\ \frac{(D+1)\alpha^2+(D-1)^2\alpha+(D^3-2D^2+5D)}{3D^3-3D^2+9D-1}\right),$$

$$a^{(3)} = \frac{1}{3D^2+1}(-\alpha^2+2D\alpha-(D^2-1),(D+1)\alpha^2+(D-1)^2\alpha+(D-1)(D+1)^2),$$

$$a^{(4)} = (\alpha+D-1,\ \alpha^2+D\alpha + (D^2+1)),$$

$$a^{(5)} = \frac{1}{3D}(-2\alpha^2+D\alpha+4D^2,\ \alpha^2+D\alpha+D^2),$$

$$a^{(6)} = \left(\frac{-(D-3)\alpha^2+2(D^2+1)\alpha-(D^3-D-4)}{3D^2+9D-8},\ \frac{(D+2)\alpha^2+(D^2-D+4)\alpha+(D^3+2D^2+5D}{3D^2+9D-8}\right),$$

$$a^{(7)} = \left(\frac{-\alpha^2+2D\alpha+2(D^2+1)}{3D^2+1},\ \frac{(D+1)\alpha^2+(D-1)^2\alpha+(D^3-2D^2-D-2)}{3D^2+1}\right).$$

Along the methods outlined for the proof of Theorem 6, the reader will by now have acquired the necessary technique to verify

THEOREM 7. Let α be the cubic irrational

$$\alpha = \sqrt[3]{D^3 + 6D}, \quad D = 2k, \quad k = 2,3,\ldots .$$

Then the JPA with $f(a^{(k)}) = \left[a^{(k)}\right]$ of $a^{(0)} = (\alpha, \alpha^2)$ is periodic; the length of the primitive pre-period is four, and the pre-period has the form

$$b^{(0)}=(2k,4k^2+3);\ b^{(1)}=(k-1,k);\ b^{(2)}=(0,1);\ b^{(3)}=(1,2k-1). \tag{4.85}$$

The primitive period has the length eight, and the period has the form

$$\begin{aligned} &b^{(4)}=(2k-1,3k^2+1);b^{(5)}=(2k-1,2k);b^{(6)}=(0,1);b^{(7)}=(0,k);\\ &b^{(8)}=(4k-1,21k^2+7);b^{(9)}=(k-1,k);b^{(10)}=(0,1);b^{(11)}=(3,2k-3). \end{aligned} \tag{4.86}$$

This result of the author was published in [2,f].

Chapter 5

VARIOUS T-FUNCTIONS

§ 1. The Inner T-Function

In this chapter various T-functions, different from $f(a^{(k)}) = \left[a^{(k)}\right]$, will be investigated. The special choice of the new T-functions described here, is motivated by the purpose that the JPA of $a^{(0)} \in E_{n-1}$ with this associated T-function becomes periodic; and periodicity of the JPA is motivated by the powerful applicational possibilities of such a JPA. The specification of such T-functions other than the conventional $f(a^{(k)}) = \left[a^{(k)}\right]$ is an advantageous novelty in the theory of these algorithms.

The most powerful T-function, so far introduced by the author, is described by

Definition IIX. Let $a^{(k)} \in E_{n-1}$ be vectors whose components are all algebraic irrationals, in symbols

$$(5.1)\quad \begin{cases} a^{(k)} = a^{(k)}(w) \equiv (a_1^{(k)}(w), a_2^{(k)}(w), \dots, a_{n-1}^{(k)}(w)) \\ w \text{ an algebraic irrational of degree } \leqslant n, \\ a^{(k)}(w) \text{ a polynomial in } w \text{ over } Q. \quad (k=0,1,\dots). \end{cases}$$

By an "inner T-function" we mean

$$(5.2)\qquad f(a^{(k)}) = \left[a^{(k)}\right]' \equiv \left(a_1^{(k)}\left([w]'\right), a_2^{(k)}\left([w]'\right), \dots, a_{n-1}^{(k)}\left([w]'\right)\right)$$

with the notation

$$(5.3)\qquad [w]' = \begin{cases} [w], & \text{if } w \geqslant 0; \\ [w] + 1, & \text{if } w < 0. \end{cases}$$

The reader should well note that the T-function $\left[a^{(k)}\right]'$ is a special case of the inner T-function. The shift of brackets from the function to the argument explains the notation of inner T-function. In the case of the JPA of the vector $a^{(0)}$ from (3.9) the function $\left[a^{(k)}\right]$ was always identical with the corresponding inner T-function. In complete analogy with P-polynomials of first order, we now introduce P-polynomials of the second order defined by

$$(5.4)\quad \begin{cases} F(x) = x^n + k_1 x^{n-1} + k_2 x^{n02} + \cdots + k_{n-1}x - d \qquad (n \geqslant 2) \\ \text{(i)} \quad k_j \quad (j=1,\ldots,n-1), \quad d \ \text{ rational integers,} \\ \text{(ii)} \quad d \neq 0, \quad d \mid k_j \quad (j=1,\ldots,n-1) \\ \text{(iii)} \quad |k_{n-1}| \geqslant c|d|(2+B); \quad E = \sum_{i=0}^{n-2} |k_i|; \quad k_0 = 1; c \geqslant 1. \end{cases}$$

We shall demonstrate the vigor of the inner T-function on P-polynomials of second order. As the reader will have guessed from the moderation of the restrictions on the coefficients of the P-polynomial by (5.4), the components of the $b^{(k)}$ vectors, appearing in the JPA with an inner T-function, can also be negative integers, and this is the great advantage of this T-function. This may look as a mere formality motivated by ambitious generalization; but since, even in this case, the convergence of the JPA, both regular and ideal, are preserved, the inner T-function will well justify its outstanding position in the theory of the JPA. The basic properties of the JPA associated with an inner T-function are stated in

THEOREM 8. A P-polynomial $F(x)$ of the second order has the following properties

(i) $F(x)$ has one and only one real root in the open interval

$$(5.5)\quad \begin{aligned} &0 < w < \frac{2}{B+4} \quad \text{for } \frac{k_{n-1}}{d} > 0, \\ &-\frac{2}{B+4} < w < 0 \quad \text{for } \frac{k_{n-1}}{d} < 0; \qquad (B \text{ as in } (5.4)) \end{aligned}$$

(ii) the JPA with the inner T-function of the vector

$$(5.6)\quad a^{(0)} = \left(\ldots, \sum_{i=0}^{s} k_i w^{s-i}, \ldots \right) \quad (s=1,\ldots,n-1)$$

is purely periodic: The length of the primitive period is n for $d \neq 1$, and 1 for $d = 1$; the period has the same form as (3.11) of Theorem 3;

(iii) the JPA with the inner T-function of the vector (5.6) is convergent, and the formula holds

(5.7)
$$w = \frac{A_0^{(sn)}}{A_0^{(sn+1)}} + \varepsilon, \qquad |\varepsilon| < r^{sn-n},$$
$$r = \left(\frac{B}{B+4}\right)^{\frac{1}{n}}, \qquad s > 1;$$

(5.8) (iv) for $c \geqslant \frac{2B}{B+2}$, $B \geqslant 2$, the JPA with the inner T-function of $a^{(0)}$ is ideally convergent.

Proof. To prove (i) we first presume

$$d \geqslant 1;$$

let $k_{n-1} > 0$; then

$$\begin{aligned} F(0) &= -d < 0, \\ F(1) &= k_{n-1} - d + 1 + k_1 + \cdots + k_{n-2} \\ &\geqslant k_{n-1} - d - (1 + |k_1| + \cdots + |k_{n-2}|) \\ &\geqslant c\, d\, (B+2) - d - B \\ &\geqslant d(B+2) - d - B = d(B+1) - B \\ &\geqslant (B+1) - B = 1 > 0, \end{aligned}$$

so that, with $F(0) < 0$, $F(1) > 0$, $F(x)$ has real roots in the interval $(0;1)$. Let further be $k_{n-1} < 0$; then

$$\begin{aligned} F(-1) &= -k_{n-1} - d \pm (1 + k_1 + k_2 + \cdots + k_{n-2}) \\ &\geqslant |k_{n-1}| - d - (1 + |k_1| + \cdots + |k_{n-2}|) \\ &\geqslant d(B+2) - d - B > 1, \end{aligned}$$

so that with $F(0) < 0$, $F(-1) > 0$, $F(x)$ has real roots in the interval $(-1;0)$. The case $d \leqslant -1$ is dealt with in complete analogy. We thus obtained

(5.9) $$0 < |w| < 1.$$

Since $F(w) = 0$, we have

(5.10) $$w = \frac{d}{k_{n-1} + k_{n-2}w + \cdots + k_1 w^{n-1} + w^n},$$

$$|w| \leqslant \frac{|d|}{|k_{n-1}| - \left(|k_{n-2}||w| + \cdots + |k_1||w|^{n-1} + |w|^n\right)},$$

so that, in view of (5.9)

$$|w| < \frac{|d|}{|k_{n-1}| - \left(|k_{n-2}| + \cdots + |k_1| + 1\right)}$$

$$\leqslant \frac{d}{|d|(B+2) - B} \leqslant \frac{d}{d\,(B+2) - d\,B} = \frac{1}{2}.$$

(5.11) $$|w| < \frac{1}{2}.$$

We also obtain from (5.9), (5.10), (5.11)

$$|w| \leqslant \frac{|d|}{|k_{n-1}| - |w|\left(|k_{n-2}| + \cdots + |k_1| + 1\right)}$$

$$\leqslant \frac{|d|}{|d|(B+2) - \frac{1}{2}B} \leqslant \frac{|d|}{|d|(B+2) - \frac{1}{2}|d|B} = \frac{2}{B+4},$$

(5.12) $$|w| < \frac{2}{B+4}.$$

For $F(x)$ from (5.4) we further obtain

$$F'(x) = nx^{n-1} + k_{n-1} + \sum_{i=1}^{n-2}(n-i)k_i x^{n-1-i},$$

$$|F'(x)| \geqslant |k_{n-1}| - n|x|^{n-1} - \sum_{i=1}^{n-2}(n-i)|k_i||x|^{n-1-i},$$

and, since we need only to investigate the case $0 < |x| < \frac{1}{2}$,

$$|F'(x)| \geqslant |k_{n-1}| - \frac{n}{2^{n-1}} - \sum_{i=1}^{n-2} \frac{n-i}{2^{n-1-i}} |k_i| .$$

But, as is known, $\frac{s}{2^{s-1}} < 1$ for $s \geqslant 2$; therefore,

$$|F'(x)| \geqslant |k_{n-1}| - 1 - \sum_{i=1}^{n-2} |k_i| = |k_{n-1}| - \sum_{i=0}^{n-2} |k_i|$$

$$= |k_{n-1}| - B \geqslant B + 2 - B > 0.$$

Since $F'(x)$ cannot vanish in the interval $0 < |x| < \frac{2}{B+4}$, (i) is completely proved.

To prove (ii), we shall carry out the JPA with the inner T-function of

$$a^{(0)}=(w+k_1,w^2+k_1w+k_2,\ldots,w^{n-1}+k_1w^{n-2}+\cdots+k_{n-2}w+k_{n-1}).$$

Since $0 < |w| < 1$ we have, both for $w > 0$ or for $w < 0$

$$[w]' = 0, \tag{5.13}$$

so that, on basis of (5.2) and (5.13)

$$b^{(0)} = (k_1,k_2,\ldots,k_{n-1}). \tag{5.14}$$

From (5.6) and (5.14) we now obtain

$$a^{(1)} = (a_1^{(0)},a_2^{(0)},\ldots,a_{n-2}^{(0)},d^{-1}a_{n-1}^{(0)}) \tag{5.15}$$

and from (5.15), again in view of (5.13) and using the inner T-function

$$b^{(1)} = (k_1,k_2,\ldots,k_{n-2},k'_{n-1}). \tag{5.16}$$

The continuation of the proof now follows faithfully the methods of proving Theorem 3. We prove by induction

$$\begin{cases} a^{(s)}=(a_1^{(0)},a_2^{(0)},\ldots,a_{n-s-1}^{(0)},d^{-1}a_{n-s}^{(0)},\ldots,d^{-1}a_{n-1}^{(0)}), \\ a^{(n-1)}=(d^{-1}a_1^{(0)},d^{-1}a_2^{(0)},\ldots,d^{-1}a_{n-1}^{(0)}), \\ a^{(n)}=(a_1^{(0)},a_2^{(0)},\ldots,a_{n-1}^{(0)}), \qquad (s=1,\ldots,n-2) \end{cases} \tag{5.17}$$

whence, always on the basis of (5.13) and using the inner T-function,

$$\begin{aligned} &b^{(s)}=(k_1,k_2,\ldots,k_{n-s-1},k'_{n-s},\ldots,k'_{n-1}), \\ &b^{(n-1)}=(k'_1,k'_2,\ldots,k'_{n-1}), \qquad (s=1,\ldots,n-2) \end{aligned} \tag{5.18}$$

which completes the proof of (ii) of Theorem 8.

The proof of (iii) of Theorem 8 follows again the lines of proving Theorem 3. We first prove

(5.19) $$\left|A_0^{(v+1)}\right| \geqslant 2\left|A_0^{(v)}\right| . \qquad (v=1,2,\ldots)$$

Since $A_0^{(1)} = A_0^{(2)} = \cdots = A_0^{(n-1)} = 0$, $A_0^{(n)} = 1$, formula (5.19) is correct for $v = 1,\ldots,n-1$. We further obtain

$$\left|A_0^{(n+1)}\right| = \left|d^{-1}\right|\left|k_{n-1}\right|A_0^{(n)}, \quad \left|A_0^{(n+1)}\right| \geqslant \left|d^{-1}\right|\left|k_{n-1}\right|$$
$$\geqslant \left|d^{-1}\right|\left|d\right|(B+2) = B+2 > 2 = 2\left|A_0^{(n)}\right|.$$

Thus (5.19) holds for $v = 1,\ldots,n$; let it be correct for $v = t,t+1,\ldots,t+n-1$; then

$$\left|A_0^{(t+n)}\right| = \left|\sum_{i=0}^{n} b^{(t)} A_0^{(t+i)}\right| =$$

$$\left|b_{n-1}^{(t)}A_0^{(t+n-1)} + A_0^{(t)} + b_1^{(t)}A_0^{(t+1)} + \cdots + b_{n-2}^{(t)}A_0^{(t+n-2)}\right| \geqslant$$

$$\left|b_{n-1}^{(t)}A_0^{(t+n-1)}\right| - \left(\left|A_0^{(t)}\right| + \left|b_1^{(t)}\right|\left|A_0^{(t+1)}\right| + \cdots + \left|b_{n-2}^{(t)}\right|\left|A_0^{(t+n-2)}\right|\right) \geqslant$$

$$\left|d^{-1}\right|\left|k_{n-1}\right|\left|A_0^{(t+n-1)}\right| - \left(\left|A_0^{(t)}\right| + \left|k_1\right|\left|A_0^{(t+1)}\right| + \cdots + \left|k_{n-2}\right|A_0^{(t+n-2)}\right) .$$

Now, by presumption, $\left|A_0^{(t+i)}\right| \geqslant \frac{1}{2}\left|A_0^{(t+1-i)}\right| \geqslant \frac{1}{2^2}\left|A_0^{(t+i-2)}\right| \geqslant \cdots \geqslant \frac{1}{2^i}\left|A_0^{(t)}\right|$, for $i=1,\ldots,n$; therefore

$$\left|A_0^{(t+n)}\right| \geqslant \left|d^{-1}\right|\left|k_{n-1}\right|\left|A_0^{(t+n-1)}\right|$$
$$-\left|A_0^{(t+n-1)}\right|\left(\frac{1}{2^{n-1}} + \frac{|k_1|}{2^{n-2}} + \cdots + \frac{|k_{n-2}|}{2}\right)$$
$$\geqslant \left|d^{-1}\right|\left|k_{n-1}\right|\left|A_0^{(t+n-1)}\right| - \left|A_0^{(t+n-1)}\right|\left(1+|k_1|+\cdots+|k_{n-2}|\right)$$
$$\geqslant (B+2)\left|A_0^{(t+n-1)}\right| - \left|A_0^{(t+n-1)}\right|B = 2\left|A_0^{(t+n-1)}\right|$$

which proves (5.19) by induction.

Denoting, as in the proof of Theorem 3,

$$H_{iv} = a_i^{(0)} - \frac{A_i^{(v)}}{A_0^{(v)}} \qquad (i=1,\ldots,n-1;\ v=n,n+1,\ldots)$$

$$M_{i,v} = \max(|H_{i,v}|,|H_{i,v+1}|,\ldots,|H_{i,v+n-2}|)$$

we obtain, as before,

$$|H_{i,v+n-1}| \leqslant \frac{\left(|A_0^{(v)}| + \sum_{j=1}^{n-2} |a_j^{(0)}||A_0^{(v+j)}|\right) M_{i,v}}{|d^{-1}a_{n-1}^{(0)}||A_0^{(v+n-1)}|},$$

whence, after easy rearrangements,

$$|H_{i,v+n-1}| \leqslant \frac{B}{B+4} M_{i,v},$$

$$|H_{i,v}| \leqslant r^{v-n}, \qquad r^n = \frac{B}{B+4}.$$

Putting $v = sn$ and recalling that $A_0^{(sn)} = k_1 A_0^{(sn-1)}$, we obtain

$$\left| a_1^{(0)} - \frac{A_1^{(sn)}}{A_0^{(sn)}} \right| < r^{sn-n},$$

$$\left| w + k_1 - \frac{k_1 A_0^{(sn} + A_0^{(sn-1)}}{A_0^{(sn)}} \right| < r^{sn-n},$$

$$\left| w - \frac{A_0^{(sn-1)}}{A_0^{(sn)}} \right| < r^{sn-n} = \left(\frac{B}{B+4}\right)^{s-1},$$

which proves (iii) of Theorem 8. (iv) of Theorem 8 is proved in complete analogy with the methods used in proving (iii) of Theorem 3.

Corollary 1 (to Theorem 8). A second order P-polynomial with $|k_{n-1}| \geqslant 2|d|B$, $B \geqslant 2$, is irreducible over Q.

Proof. This follows from the specification of c from (iv) of Theorem 8 and from the ideal convergence of the JPA with the inner T-function of $a^{(0)}$.

Corollary 2 (to Theorem 8). Statements (i), (ii), (iii) of Theorem 8 remain valid, if the restrictions on the coefficients of the second order P-polynomial, viz. $d|k_i$ $(i=1,\dots,n-1)$ are removed.

The reader will have no difficulty to verify this Corollary. For it is obvious from the proofs of these statements that the property $d|k_i$ is never made use of. But it should be emphasized that statement (iv) of Theorem 8 does not hold without the property $d|k_i$. The situation $d \nmid k_i$ will be investigated in the next paragraph. It is also obvious from this Corollary that the components of the vectors $b^{(k)}$ are not all integers in this case. For number-theoretic considerations this may look disadvantageous, but for algebraic needs the inner T-function is the more powerful.

§2. Irreducibility of Polynomials

Before we state our main irreducibility criterion for polynomials, we shall still investigate those special cases which were excluded in §1. The reader will recall that for the irreducibility of the second order P-polynomial $F(x) = x^n + k_1x^{n-1} + \cdots + k_{n-1}x - d$ with $d|k_i$ $(i=1,\ldots,n-1)$, $B = \sum_{i=0}^{n-2} |k_i|$, the condition $|k_{n-1}| \geqslant 2|d|B$, $B \geqslant 2$ must be fulfilled. We shall now take a better look at the case

$$B = 1;\ k_1 = k_2 = \cdots = k_{n-2} = 0; \tag{5.20}$$
$$F(x) = x^n + k_{n-1}x - d;\quad d|k_{n-1}.$$

THEOREM 9. Let F(x) be a second order polynomial of the form (5.20). If

$$|k_{n-1}| \geqslant 3|d| \tag{5.21}$$

then F(x) is irreducible over Q.

Proof. From (5.5) we obtain, putting B = 1,

$$0 < |w| < \tfrac{2}{5},\quad w^n + k_{n-1}w - d = 0, \tag{5.22}$$

where again w is the unique root of F(x) in the open interval (-1,0) if $d^{-1}k_{n-1} < 0$, and (0,1) if $d^{-1}k_{n-1} > 0$. With the notations

$$L_{i,v} = A_i^{(v)} - a_i^{(0)} A_0^{(v)}, \quad N_{i,v} = \max(|L_{i,v}|, \ldots, |L_{i,v+n-2}|)$$

we obtain, as before,

$$|L_{i,v+n-1}| \leqslant q\left(1 + \sum_{j=1}^{n-2} |a_j^{(0)}|\right) N_{i,v}, \quad q = |w|. \quad (v \geqslant 1)$$

Now $1 + \sum_{j=1}^{n-2} |a_j^{(0)}| = 1 + \sum_{j=1}^{n-2} q^j < 1 + \sum_{j=1}^{\infty} q^j$

$$= \frac{1}{1-q} < \frac{1}{1-\frac{2}{5}} = \frac{5}{3},$$

$|L_{i,v+n-1}| < \frac{2}{3} N_{i,v}$, and from here, as before,

$|A_i^{(v)} - a_i^{(0)} A_0^{(v)}| < \left(\frac{2}{3}\right)^{\left[\frac{v}{n}\right] - 1}$, $(v \geqslant n)$ which proves ideal convergence of the JPA with the inner T-function of the vector $a^{(0)} = (w, w^2, \ldots, w^{n-2}, w^{n-1} + k_{n-1})$ and with it the irreducibility of the $F(x) = x^n + k_{n-1}x - d$.

We can state Theorem 9, in the wording that the polynomial

(5.23) $F(x) = x^n + 3dtx - d$; d,t rational integers $\neq 0$

is irreducible over Q.

THEOREM 10. The first order P-polynomial

(5.24) $F(x) = x^n + 2x - 1 \quad (n \geqslant 2)$

is irreducible over Q.

Proof. It is immediately obvious that F(x) has one and only one real root w in the open interval $0 < w < \frac{1}{2}$. As before, we obtain

$$|L_{i,v+n-1}| \leqslant w\left(1 + \sum_{j=1}^{n-2} |a_j^{(0)}|\right) N_{i,v}$$

$$= w\left(1 + \sum_{j=1}^{n-2} w^j\right) N_{i,v} = \frac{w(1 - w^{n-1})}{1 - w} N_{i,v}$$

$$= \frac{w - w^n}{1 - w} N_{i,v};$$

but $w^n = 1 - 2w$; therefore

$$|L_{i,v+n-1}| \leqslant \frac{3w - 1}{1 - w} N_{i,v}.$$

But $(3w - 1)/(1 - w) < 1$, since $w < \frac{1}{2}$; thus

$$\left|A_i^{(v)} - a_i^{(0)} A_0^{(v)}\right| < \left(\frac{3w - 1}{1 - w}\right)^{\left[\frac{v}{n}\right] - 1}, \quad (v \geqslant n)$$

which proves ideal convergence, and therefore irreducibility of $F(x)$.

THEOREM 11. The polynomial

$$F(x) = x^n + x - 1 \tag{5.25}$$

is irreducible over Q.

Proof. Here we shall need some new technique. It is obvious that $F(x)$ has one and only one real root w in the open interval $(0,1)$. From $w^n + w - 1 = 0$, $w^n = 1 - w$, we further obtain

$$w = \frac{1}{1 - w^{n-1}} < \frac{1}{1 + w^n} = \frac{1}{2 - w},$$

$$2w - w^2 < 1, \quad (w - 1)^2 > 0,$$

which is correct for any $w \neq 1$. We can therefore take any upper bound for w; but there is a lower bound, viz.

$$w = \frac{1}{1 + w^{n-1}} \geqslant \frac{1}{1 + w}; \quad w^2 + w > 1, \quad \text{for } n \geqslant 3,$$

$w^2 + w - 1 > 0$, $w > \dfrac{\sqrt{5} - 1}{2}$, and we can take

(5.26) $$\frac{\sqrt{5}-1}{2} < w < \frac{2}{3}.$$

We now obtain, as in Theorem 10,

$$|L_{i,v+n-1}| \leqslant \frac{w - w^n}{1 - w} N_{i,v} = \frac{2w - 1}{1 - w} N_{i,v}.$$

But $\frac{2w - 1}{1} < 1$ for $w < \frac{2}{3}$, and $\frac{2w - 1}{1 - w} > 0$ for $w > \frac{\sqrt{5}-1}{2}$.

Hence

$$|A_i^{(v)} - a_i^{(0)} A_0^{(v)}| \leqslant \left(\frac{2w - 1}{1 - w}\right)^{[v/n] - 1}. \qquad (v \geqslant n)$$

We shall now advance to prove the main result of this section which is stated in

THEOREM 12. The polynomial

(5.27) $$\begin{cases} F(x) = x^n + k_1 x^{n-1} + \cdots + k_{n-1} x - d, \\ k_i \ (i = 1,\ldots,n-1),\ d \text{ rational integers}, \\ |k_{n-1}| \geqslant 2B|d|^{n+1}; \quad |d| \geqslant 2; \quad n \geqslant 2; \\ B = \sum_{i=0}^{n-2} |k_i|; \quad k_0 = 1 \end{cases}$$

is irreducible over Q.

Proof. The reader will first note that the coefficients of the polynomial in question are not restricted any more by the condition $d|k_i$ $(i = 1,\ldots,n-1)$, as was the case with the first and second order P-polynomials. The case $|d| = 1$ is excluded, since it belongs to P-polynomials.

For $|k_{n-1}| \geqslant 2B|d|^{n+1}$, we obtain

$$|k_{n-1}| \geqslant 8B|d| \geqslant c|d|(2 + B) \quad \text{for } c > 2,$$

so that condition (iii) of a second order P-polynomial is fulfilled and F(x) of (5.27) has one and only one root w in the interval (0,1) or (-1,0), so that

(5.28) $$0 < |w| = q < 1.$$

We further recall that the JPA with the inner T-function of the vector

$$a^{(0)} = (w+k_1, w^2+k_1w+k_2, \dots, w^{n-1}+k_1w^{n-2}+\dots+k_{n-1})$$

is purely periodic with length of the primitive period = n; the period has the form (3.11); the only difference in the present case is, that the numbers $d^{-1}|k_i$ (i = 1,...,n-1) are not any more integers. We further obtain

$$q \leqslant \frac{|d|}{|k_{n-1}| - (q^{n-1} + |k_1|q^{n-2}+\dots+|k_{n-2}|q)}$$

$$< \frac{|d|}{|k_{n-1}| - (1 + |k_1| +\dots+ |k_{n-2}|)}$$

$$\leqslant \frac{|d|}{2B|d|^{n+1} - B} < \frac{|d|}{2B|d|^{n+1} - B|d|^{n+1}};$$

$$q < \frac{1}{B|d|^n};$$

$$q \leqslant \frac{|d|}{|k_{n-1}| - q(q^{n-2} + |k_1|q^{n-3}+\dots+ |k_{n-2}|)}$$

$$< \frac{|d|}{|k_{n-1}| - q(1 + |k_1| +\dots+ |k_{n-2}|)}$$

$$\leqslant \frac{|d|}{2B|d|^{n+1} - qB} < \frac{|d|}{2B|d|^{n+1} - \frac{1}{|d|^n}} < \frac{|d|}{2B|d|^{n+1} - |d|},$$

(5.29) $$0 < q < \frac{1}{2B|d|^n - 1}.$$

We make use again of the formula

$$|L_{i,v+n-1}| \leqslant q\left(1+\sum_{j=1}^{n-2}|a_j^{(0)}|\right)N_{i,v}$$

$$\leqslant q\left(1+\sum_{j=1}^{n-2}(q^j+|k_1|q^{j-1}+\cdots+|k_{j-1}|q+|k_j|)\right)N_{i,v}$$

$$= q\left(1+q+\cdots+q^{n-2}+\sum_{j=1}^{n-2}(1+q+\cdots+q^{n-2-j})|k_j|\right)N_{i,v}$$

$$= \frac{q}{1-q}\left[1-q^{n-1}+\sum_{j=1}^{n-2}(1-q^{n-1-j})|k_j|\right]N_{i,v}$$

$$\leqslant \frac{q}{1-q}\left[1-q^{n-1}+\sum_{j=1}^{n-2}(1-q^{n-1})|k_j|\right]N_{i,v}$$

$$= \frac{qB}{1-q}(1-q^{n-1})N_{i,v},$$

(5.30) $$|L_{i,v+n-1}| \leqslant \frac{qB}{1-q}(1-q^{n-1})N_{i,v}. \qquad (v \geqslant 1)$$

In virtue of (5.29) we obtain

$$\frac{qB}{1-q} < \frac{B}{2B|d|^n-2} < \frac{B}{2B|d|^n-B|d|^n} = \frac{1}{|d|^n},$$

$$L_{i,v+n-1} \leqslant \frac{1-q^{n-1}}{|d|^n}N_{i,v},$$

hence, as before,

(5.31) $$\left|A_i^{(v)}-a_i^{(0)}A_0^{(v)}\right| < \left(\frac{r}{|d|}\right)^v, \qquad r=\sqrt[n]{1-q^{n-1}};\ v \geqslant n.$$

Decisive for the proof of Theorem 12 is

Lemma 2. The numbers

(5.32) $$d^{(s-1)(n-1)+t}A_i^{(sn+t)} \qquad (s=1,2,\ldots;i,t=0,\ldots,n-1)$$

are rational integers.

The reader will recall that, since the components of the vectors $b^{(v)}$ appearing in the JPA with the inner T-function of $a^{(0)}$ are not always integers, also the $A_i^{(v)}$ $(i=0,\ldots,n-1;\ v=n,n+1,\ldots)$ are not always integers. We prove that the numbers (5.32) are. Proof is by induction. We calculate easily:

$$A_i^{(n)} = k_i;\ A_i^{(n+1)} = k_{i-1} + d^{-1}k_ik_{n-1},$$

$$\text{(5.33)}\qquad dA_i^{(n+1)} = B_i^{(n+1)};\quad B_i^{(n+1)} \text{ a rational integer.}$$

We prove the formula

$$\text{(5.34)}\qquad d^jA_i^{(n+j)} = B_i^{(n+j)};\quad B_i^{(n+j)} \text{ a rational integer.}$$

$$(i,t=0,\ldots,n-1)$$

Formula (5.34) is true for $j=0,1$. Presume its correctness for $j=0,1,\ldots,t$ $(t=0,\ldots,n-2)$

$$d^{t+1}A_i^{(n+t+1)} = d^{t+1}\Big(A_i^{(t+1)}+k_1A_i^{(t+2)}+\cdots+k_{n-t-2}A_i^{(n-1)}$$

$$+ d^{-1}k_{n-t-1}A_i^{(n)}+d^{-1}k_{n-t}A_i^{(n+1)}+\cdots+d^{-1}k_{n-1}A_i^{(n+t)}\Big);$$

but $A_i^{(t+s)} = 0$, for $t+s \neq i$, $A_i^{(t+s)} = 1$, for $t+s=i$, $t+1 \leqslant t+s \leqslant n-1$; in any case $A_i^{(t+1)}+k_1A_i^{(t+2)}+\cdots+k_{n-t-2}A_i^{(n-1)} = D$, D a rational integer; hence $d^{t+1}A_i^{(n+t+1)} = d^{t+1}D + d^tk_{n-t-1}A_i^{(n)}+d^tk_{n-t}A_i^{(n+1)}+\cdots$ $+d^tk_{n-1}A_i^{(n+t)}$. But $d^tA_i^{(n)}$, $d^tA_i^{(n+1)},\ldots,d^tA_i^{(n+t)}$ are all rational integers by presumption, which proves (5.34). We further calculate

$$d^{n-1}A_i^{(2n)} = d^{n-1}\left(A_i^{(n)} + \sum_{j=1}^{n-1} k_jA_i^{(n+j)}\right)$$

$$= d^{n-1}k_i + \sum_{j=1}^{n-1} k_jd^{n-1-j}\ d^jA_i^{(n+j)}$$

$$= d^{n-1}k_i + \sum_{j=1}^{n-1} k_jd^{n-1-j}\ B_i^{(n+j)},$$

$$\text{(5.35)}\qquad d^{n-1}A_i^{(2n)} = B_i^{(2n)},\quad B_i^{(2n)} \text{ a rational integer.}$$

The following formula is now easily proved in the same way as (5.34).

(5.36) $d^{n-1+j}A_i^{(2n+j)} = B_i^{(2n+j)}$; $B_i^{(2n+j)}$ rational integers;

(j=0,...,n-1).

We now obtain, in virtue of (5.35), (5.36)

$$d^{2(n-1)}A_i^{(3n)} = d^{2(n-1)}\left(A_i^{(2n)} + \sum_{j=1}^{n-1} k_j A_i^{(2n+j)}\right)$$

$$= d^{2(n-1)}A_i^{(2n)} + \sum_{j=1}^{n-1} k_j d^{n-1-j} d^{n-1+j} A_i^{(2n+j)}$$

$$= d^{n-1}B_i^{(2n)} + \sum_{j=1}^{n-1} k_j d^{n-1-j} B_i^{(2n+j)},$$

(5.37) $d^{2(n-1)}A_i^{(3n)} = B_i^{(3n)}$; $B_i^{(3n)}$ a rational integer.

(i=0,...,n-1)

Using this technique, the reader will see that Lemma 2 is now proved by induction.

We shall return to formula (1.14) to obtain

$$\pm\left[A_0^{((s+1)n+t-1)} + \sum_{j=1}^{n-1} a_j^{(v)} A_0^{((s+1)n+t-1+j)}\right]^{-1} =$$

$$\begin{vmatrix} A_1^{((s+1)n+t)} - a_1^{(0)} A_0^{((s+1)n+t)}, \ldots, A_1^{((s+1)n+t+n-2)} - a_1^{(0)} A_0^{((s+1)n+t+n-2)} \\ A_2^{((s+1)n+t)} - a_2^{(0)} A_0^{((s+1)n+t)}, \ldots, A_2^{((s+1)n+t+n-2)} - a_2^{(0)} A_0^{((s+1)n+t+n-2)} \\ \cdots\cdots\cdots\cdots\cdots\cdots\cdots\cdots\cdots\cdots\cdots\cdots \\ A_{n-1}^{((s+1)n+t)} - a_{n-1}^{(0)} A_0^{((s+1)n+t)}, \ldots, A_{n-1}^{((s+1)n+t+n-2)} - a_{n-1}^{(0)} A_0^{((s+1)n+t+n-2)} \end{vmatrix}.$$

We now multiply the first column of this determinant by $d^{s(n-1)+t}$, the second by $d^{s(n-1)+t+1}$,..., the last by $d^{s(n-1)+t+n-2}$ and obtain in virtue of (5.32) and putting $v = (s+1)n+t-1$

$$(5.38)\quad \begin{cases} \pm\, d^{\frac{1}{2}(n-1)(2v+n)} \left[\sum_{j=0}^{n-1} a_j^{(v)} A_0^{(v+j)} \right]^{-1} = \\[2ex] \begin{vmatrix} B_1^{(v+1)} - a_1^{(0)} B_0^{(v+1)}, \ldots, B_1^{(v+n-1)} - a_1^{(0)} B_0^{(v+n-1)} \\ B_2^{(v+1)} - a_2^{(0)} B_0^{(v+1)}, \ldots, B_2^{(v+n-1)} - a_2^{(0)} B_0^{(v+n-1)} \\ \cdots\cdots\cdots\cdots\cdots\cdots \\ B_{n-1}^{(v+1)} - a_{n-1}^{(0)} B_0^{(v+1)}, \ldots, B_{n-1}^{(v+n-1)} - a_{n-1}^{(0)} B_0^{(v+n-1)} \end{vmatrix} . \end{cases}$$

On the basis of previous considerations, we now see that Theorem 12 will be proved if we can show that the sequences $\left\langle B_i^{(v)} - a_i^{(0)} B_0^{(v)} \right\rangle$ $(i=1,\ldots,n-1)$ are all null sequences. This is easily done. We have, by definition, for $v = (s+1)n+t-1$,

$$\left| B_i^{(v)} - a_i^{(0)} B_0^{(v)} \right| = \left| B_i^{((s+1)n+t-1)} - a_i^{(0)} B_0^{((s+1)n+t-1)} \right|$$

$$= |d|^{s(n-1)+t-1} \left| A_i^{((s+1)n+t-1} - a_i^{(0)} A_0^{((s+1)n+t-1)} \right| ,$$

hence, by (5.31),

$$\left| B_i^{(v)} - a_i^{(0)} B_0^{(v)} \right| < |d|^{s(n-1)+t-1} \cdot \left(\frac{r}{|d|} \right)^{(s+1)n+t-1}$$

$$= \frac{r^{(s+1)n+t-1}}{|d|^{n+s}} \leqslant \frac{r^v}{|d|^{\left[\frac{v}{n}\right]}} \leqslant \left(\frac{r}{|d|} \right)^{\left[\frac{v}{n}\right]} . \qquad (v \geqslant n)$$

Since $r < 1$, $|d| \geqslant 2$, $\left| B_i^{(v)} - a_i^{(0)} B_0^{(v)} \right|$ is a null sequence.

There exist numerous other irreducibility criteria for polynomials. But these are mainly based on divisibility relations, while the criterion, as stated in Theorem 12, is based on magnitude of the coefficients. For $n = 2$, this criterion maintains (since in this case $B = 1$) that the quadratic polynomial $x^2 + k_1 x - d$ is irreducible for $|k_1| \geqslant 2|d|^3$. Indeed, we obtain, denoting the discriminant by Δ,

$$(|k_1| - 1)^2 < |k_1|^2 - 4|d| = \Delta < |k_1|^2 + 4|d| < (|k_1| + 1)^2,$$

and since $\Delta \neq |k_1|^2$ for $d \neq 0$, that Δ is not a perfect square.

We shall conclude this paragraph with a numeric illustration of the theory developed in this chapter.

Example 8. We choose the second order P-function

$$F(x) = x^5 - 2x^3 + 4x^2 - 28x + 2.$$

Here $B = 7$; $|k_{n-1}| = 28 = 2\cdot 2\cdot 7 = 2|d|B$. On the basis of Theorem 8, $F(x)$ has one and only one real root w in the open interval $(-\frac{2}{11}, 0)$. On basis of Corollary 1 to Theorem 8, $F(w)$ is irreducible over Q, w is a fifth degree irrational. The same Theorem also states: the JPA with the inner T-function of the vector

$$a^{(0)} = (w, w^2-2, w^3-2w+4, w^4-2w^2+4w - 28)$$

is purely periodic, and its period has the form

$$b^{(0)} = (0, -2, 4, -28),$$

$$b^{(1)} = (0, -2, 4, 14),$$

$$b^{(2)} = (0, -2, -2, 14),$$

$$b^{(3)} = (0, 1, -2, 14),$$

$$b^{(4)} = (0, 1, -2, 14).$$

w can be rationally approximated by

$$\left| w - \frac{A_0^{(5s-1)}}{A_0^{(5s)}} \right| < \left(\frac{7}{11}\right)^{s-1} .$$

For $s = 2$, we calculate easily

$$A_0^{(9)} = 37272; \; A_0^{(10)} = -1.033.247$$

$$w = -\frac{37272}{1.033.247} \pm \frac{7}{11} .$$

Of course, the approximation error here is much too large.

§ 3. Application of the Inner T-function

In this paragraph we shall show how much easier life becomes, when the JPA is associated with the inner T-function; where the "outer" T-function $f(a^{(k)}) = [a^{(k)}]$ results in a periodic JPA only after much effort — and sometimes not at all, at least not as far as man or the computer can afford — the inner T-function does it! Think only about the JPA associated with $f(a^{(k)}) = [a^{(k)}]$ of $a^{(0)} = (w, w^2, \ldots, w^{n-1})$, $w = \sqrt[n]{D^n - d}$ with its period of length n^2 (and pre-period of length $(n-1)^2 + 1$; see [2,g]). It is exactly this case where we start to illustrate the power of the inner T-function. Let F(x) be a second order P-polynomial of the form

$$(5.39)\quad \begin{cases} F(x) = x^n + \binom{n}{1}Dx^{n-1} + \binom{n}{2}D^2x^{n-2} + \cdots + \binom{n}{n-1}D^{n-1}x + d, \\ \qquad d, D \text{ natural numbers}, \quad d \mid D, \; n \geqslant 2 \\ \qquad\qquad D \geqslant 2(n-1)(e-2)d. \end{cases}$$

The reader will verify easily that

$$\binom{n}{n-1}D^{n-1} = |k_{n-1}| \geqslant 2|d|B, \qquad B > 2,$$

$$B = 1 + \binom{n}{1}D + \binom{n}{2}D^2 + \cdots + \binom{n}{n-2}D^{n-2}$$

so that by Corollary 1 (to Theorem 8) F(x) is irreducible and the JPA with the inner T-function of the vector

$$(5.40)\qquad a^{(0)} = (w + \tbinom{n}{1}D, w^2 + \tbinom{n}{1}Dw + \tbinom{n}{2}D^2, \ldots, w^{n-1} + \tbinom{n}{1}Dw^{n-2} + \cdots + \tbinom{n}{n-1}D^{n-1})$$

where w is the only real root of F(x) in the open interval $(-\frac{B}{B+4}, 0)$, is periodic and ideally convergent. The reader should note that the d of the second order P-polynomial F(x) in (5.40) equals -d, and the period has the form (for $d > 1$)

$$(5.41)\quad \begin{cases} b^{(0)} = \left(\binom{n}{1}D,\ \binom{n}{2}D^2,\dots,\binom{n}{n-1}D^{n-1}\right), \\ b^{(s)} = \left(\binom{n}{1}D,\dots,\binom{n}{n-s-1}D^{n-s-1},\binom{n}{n-s}D^{n-s-1}t,\dots,\binom{n}{n-1}D^{n-2}t\right) \\ \qquad s = 1,\dots,n-2, \\ b^{(n-1)} = \left(\binom{n}{1}t,\ \binom{n}{2}Dt,\dots,\binom{n}{n-1}D^{n-2}t\right) \\ \qquad t = -d^{-1}D. \end{cases}$$

We can rearrange $F(w)$ in the form

$$(w+D)^n - D^n + d = 0, \qquad w+D = \sqrt[n]{D^n - d}$$

$$(5.42)\qquad w = \sqrt[n]{D^n - d} - D.$$

w is calculated by means of formula (5.7), and $\sqrt[n]{D^n - d}$ from (5.42). Denoting $-d^{-1} = g$, one calculates easily, on basis of (5.41)

$$A_0^{(n)} = 1;\quad A_0^{(n+1)} = \binom{n}{1}g\,D^{n-1};$$

$$A_0^{(n+2)} = \binom{n}{1}^2 g^2 D^{2n-2} + \binom{n}{2}g\,D^{n-2};$$

$$A_0^{(n+3)} = \binom{n}{1}^3 g^3 D^{3n-3} + 2\binom{n}{1}\binom{n}{2}g^2D^{2n-3} + \binom{n}{3}g\,D^{n-3};$$

$$A_0^{(n+4)} = \binom{n}{1}^4 g^4 D^{4n-4} + 3g^3\binom{n}{1}^2\binom{n}{2}D^{3n-4}$$
$$+ \left[2\binom{n}{1}\binom{n}{3}+\binom{n}{2}^2\right] g^2D^{2n-4} + \binom{n}{4}g\,D^{n-n};$$

the reader will have no difficulty in verifying that the fractions $A_0^{(v)}/A_0^{(v+1)}$ $(v=n,n+1,\dots)$ are all negative. It should also be noted that if we demand from the JPA with the inner T-function of $a^{(0)}$ from (5.40) to be only periodic, disregarding convergency, then the restriction $D \geqslant 2(n-1)(e-2)d$ can be dropped, and only $d|D$ is necessary, if we want the components of the $b^{(v)}$ to be rational integers; if even this is not demanded, then no restrictions need to be imposed on D and d.

Completely analogous considerations (as were investigated in Chapter 4, §1) hold for the first order P-polynomial

(5.43) $$F(x) = x^n+\binom{n}{1}Dx^{n-1}+\dots+\binom{n}{n-1}D^{n-1}x - d; \quad n \geqslant 2;$$

$$d, D \text{ natural numbers}; \quad d|D; \quad D \geqslant 2(n-1)(e-2)d.$$

Here $F(x)$ has one and only one real root w in the open interval $(0, \frac{B}{B+4})$; the JPA with the inner T-function of the vector

$$a^{(0)} = (w+\binom{n}{1}D, w^2+\binom{n}{1}Dw+\binom{n}{2}D^2, \dots, w^{n-1}+\binom{n}{1}Dw^{n-2}+\dots+\binom{n}{n-1}D^{n-1})$$

is ideally convergent and purely periodic, and the period has the form of (5.41); only here the components of the vectors $b^{(v)}$ are all positive integers (since $t = d^{-1}D$) and have the same form as in (4.15). We further obtain, after easy rearrangements

$$w + D = \sqrt[n]{D^n + d},$$

(5.44) $$w = \sqrt[n]{D^n + d} - D.$$

If the restriction $D \geqslant 2(n-1)(e-2)d$ is removed and only $d|D$ is maintained, the JPA of $a^{(0)}$ is still convergent; also if the restriction $d|D$ is removed. With this the question of the periodicity of a JPA involving any irrational $\sqrt[n]{m}$ ($n \geqslant 2$; $m > 0$ and not a perfect n-th power) is completely solved. Of course, the condition of integrality of the components of $b^{(v)}$ must, regrettably, be resigned from. This would hurt the integrity of a number-theoretician.

We shall now investigate the first order P-polynomial

(5.45) $$F(x) = x^3 + 3Dx^2 + 3D^2x - 3D; \quad D \in N.$$

This polynomial has one and only one real root w in the open interval $(0, \frac{1}{1+D})$ as can be easily verified. The JPA with the inner T-function of the vector

(5.46) $$a^{(0)} = (w+3D, \; w^2+3Dw + 3D^2)$$

is purely periodic and the period has the form, for any D,

$$(5.47)\qquad \begin{cases} b^{(0)} = (3D,\ 3D^2), \\ b^{(1)} = (3D,\ D), \\ b^{(2)} = (1,\ D); \end{cases}$$

this JPA is, of course, convergent in virtue of Theorem 1; but we cannot approximate w by the previous methods, since the coefficient k_{n-1} does not fulfill the conditions required there. From (5.45) we obtain, since $w^3 + 3Dw^2 + 3D^2w = 3D$,

$$(w + D)^3 = D^3 + 3D,$$

$$(5.48)\qquad \alpha = \sqrt[3]{D^3 + 3D} = w + D.$$

This is the familiar cubic irrational (4.80) dealt with by Theorem 6. But while the JPA of (α,α^2) with $f(a^{(k)}) = \left[a^{(k)}\right]$ was quite complicated, the JPA with the inner T-function of the vector

$$a^{(0)} = (\alpha + 2D,\ \alpha^2 + D\alpha + D^2)$$

uncomplicates the situation profoundly. More than that — for $D = 1$, the JPA with $f(a^{(k)}) = \left[a^{(k)}\right]$ "seems to occupy a most magic place...", while the JPA with the inner T-function of the vector

$$a^{(0)} = (\sqrt[3]{4} + 2,\ \sqrt[3]{16} + \sqrt[3]{4} + 1)$$

is purely periodic with the period

$$b^{(0)} = (3,\ 3),$$

$$b^{(1)} = (3,\ 1),$$

$$b^{(2)} = (1,\ 1).$$

This again demonstrates the great advantage of the inner T-function.

We shall now introduce a new element in the JPA of algebraic vectors. As the reader will verify easily, the components of the vectors, $a^{(0)}$ so far investigated, were all algebraic integers. That the JPA of $a^{(0)}$ can become periodic with the components of the vectors $b^{(v)}$ all being rational integers without the restriction that the components of $a^{(0)}$ be algebraic integers will be shown in the following theorem; this new technique was necessitated by the use of new irrationalities whose JPA could otherwise not be made periodic. In [2,n] the author proved:

THEOREM 13. Let denote

$$(5.49)\qquad \begin{aligned} & w = \sqrt[n]{D^n + dD}; \qquad d,D,n \in N; \\ & d \mid D;\ n-1 \leqslant d \leqslant \frac{D}{n-2};\quad n \geqslant 2 \end{aligned}$$

$$(5.50)\qquad \begin{aligned} & a_s^{(0)\,'} = \sum_{i=0}^{s} \binom{n-1-s+i}{i} w^{s-i} D^i; \qquad (s=1,\dots,n-1) \\ & a^{(0)} = \left(d^{-1}a_1^{(0)\,'},\ d^{-2}a_2^{(0)\,'},\dots,d^{-(n-1)}a_{n-1}^{(0)\,'}\right). \end{aligned}$$

Then the JPA with the T-function $f(a^{(k)}) = [a^{(k)}]$ is purely periodic with a primitive period of length n, if $D \neq d^{n-1}$ and of length 1, if $D = d^{n-1}$. Denoting $t = d^{-1}D$, the period takes the form

$$(5.51)\quad \begin{cases} b^{(0)} = \left(\binom{n}{1}t, \binom{n}{2}t^2,\dots,\binom{n}{n-1}t^{n-1}\right) \\ \begin{cases} b^{(m-1)} = \left(\binom{n}{1}t,\dots,\binom{n}{n-m}t^{n-m}, \binom{n}{n-m+1}d^{m-2}D^{n-m},\right. \\ \qquad\left.\binom{n}{n-m+2}d^{m-3}D^{n-m+1},\dots,\binom{n}{n-1}D^{n-2}\right) \quad m=2,\dots,n-1 \end{cases} \\ b^{(n-1)} = \left(\binom{n}{1}d^{n-2}, \binom{n}{2}d^{n-3}D,\dots,\binom{n}{n-2}dD^{n-3}, \binom{n}{n-1}D^{n-2}\right). \end{cases}$$

Corollary 1 (to Theorem 13). In the case $n = 3$, the restrictions on d can be removed to the only condition that $d \mid D$.

Corollary 2 (to Theorem 13). Theorem 13 remains valid, if instead of the JPA with the outer T-function $f(a^{(k)}) = [a^{(k)}]$ the inner T-function is applied; in this case only the restriction $d \mid D$ has to be preserved.

We shall illustrate Corollary 2 (to Theorem 13) in the case $n = 3$. We obtain

$$w = \sqrt[3]{D^3 + dD};\ w^3 - D^3 = dD;\ \frac{1}{w-D} = \frac{w^2+Dw+D^2}{dD};$$

$$a^{(0)} = \left(\frac{w + 2D}{d},\ \frac{w^2 + Dw + d^2}{d^2}\right)$$

$$b^{(0)} = \left(\frac{3D}{d},\ \frac{3D^2}{d^2}\right)$$

$$a^{(0)} - b^{(0)} = \left(\frac{w - D}{d},\ \frac{(w - D)(w + 2D)}{d^2}\right)$$

$$a^{(1)} = \left(\frac{w + 2D}{d},\ \frac{w^2 + wD + D^2}{D}\right)$$

$$b^{(1)} = \left(\frac{3D}{d},\ 3D\right)$$

$$a^{(1)} - b^{(1)} = \left(\frac{w - D}{d},\ \frac{(w - D)(w + 2D)}{D}\right)$$

$$a^{(2)} = \left(\frac{d(w + 2D)}{D},\ \frac{w^2 + Dw + D^2}{D}\right)$$

$$b^{(2)} = (3d,\ 3D)$$

$$a^{(2)} - b^{(2)} = \left(\frac{d(w - D)}{D},\ \frac{(w - D)(w + 2D)}{D}\right)$$

$$a^{(3)} = \left(\frac{w + 2D}{d},\ \frac{w^2 + Dw + D^2}{d^2}\right)$$

$$a^{(3)} = a^{(0)}.$$

If we substitute $t = d^{-1}D$ in the components of $b^{(0)}$ and $b^{(s)}$, we obtain

$$b^{(0)} = (3t,\ 3t^2)$$

$$b^{(1)} = (3t,\ 3D)$$

$$b^{(2)} = (3d,\ 3D)$$

which is the form given in (5.51) for $n = 3$. If $D = d^2$, then $t = d$, $D = t^2$, and the vectors $b^{(1)} = b^{(2)} = b^{(0)} = (3t,\ 3t^2)$, so that in this case the length of the primitive period is one, as demanded by Theorem 13.

§4. The Inner-Outer T-Function

As was demonstrated in the preceding paragraphs, the JPA with the outer or inner T-functions becomes periodic for the vector $a^{(0)} \in E_{n-1}$, where components are usually polynomials in a certain irrationality w of degree n. These polynomials were always arranged in ascending degrees, whereupon the JPA of $a^{(0)}$ became periodic. One could, of course, ask the question to what an extent some periodicity would be preserved (with a new form of the primitive period) if the order of these polynomials were to be rearranged. The aimfulness of doing so would be prompted not only by mathematical curiosity — new information about the behavior of the JPA associated with this or another T-function could be derived. Indeed, this question was fed into the computer and the following results were obtained, always operating with the JPA of $a^{(0)} = (w,w^2)$ or $a^{(0)} = (w^2,w)$ associated with the outer T-function $f(a^{(k)}) = \left[a^{(k)}\right]$:

(i) $a^{(0)} = (w,w^2)$; $w = \sqrt[3]{2}$;
length of pre-period 2,
length of period 1.

(ii) $a^{(0)} = (w^2,w)$; $w = \sqrt[3]{2}$;
length of pre-period 1,
length of period 2.

(iii) $a^{(0)} = (w,w^2)$; $w = \sqrt[3]{3}$;
length of pre-period 2,
length of period 2.

(iv) $a^{(0)} = (w^2,w)$; $w = \sqrt[3]{3}$;
length of pre-period 6,
length of period 17.

(v) $a^{(0)} = (w,w^2)$; $w = \sqrt[3]{10}$;
length of pre-period 2,
length of period 3.

(vi) $a^{(0)} = (w^2, w)$; $w = \sqrt[3]{10}$;

length of pre-period 10,

length of period 22.

(vii) $a^{(0)} = (w, w^2)$; $w = \sqrt[3]{18}$;

length of pre-period 2,

length of period 3.

(viii) $a^{(0)} = (w, w^2)$; $w = \sqrt[3]{18}$;

length of pre-period 8,

length of period 10.

(ix) $a^{(0)} = (w, w^2)$; $w = \sqrt[3]{14}$;

length of pre-period 4,

length of period 9.

(x) $a^{(0)} = (w^2, w)$; $w = \sqrt[3]{14}$;

length of pre-period 8,

length of period 14.

Of course, the computer showed no inclination of periodicity for the vectors (w, w^2) or (w^2, w) with $w = \sqrt[3]{4}$. The author when faced with the question of periodicity of the JPA of the vector $a^{(0)} = (w^{n-1}, w^{n-2}, \ldots, w^2, w)$, w an n-th degree irrationality constructed a new T-function in order to overcome these difficulties; these results were published in [2,h]. This new "inner-outer T-function" is of a rather complicated nature, but it has justified its application for the purpose posed above.

The inner-outer T-function is constructed in the following manner: let again the components of $a^{(v)}$ be functions of the n-th degree irrationality w; as a result of the JPA operations, the components $a_i^{(v)}$ are obtained by the division process

$$a_i^{(v)} = \frac{a_{i+1}^{(v-1)} - b_{i+1}^{(v-1)}}{a_1^{(v-1)} - b_1^{(v-1)}} . \qquad (v=1,2,\ldots)$$

To find $b_i^{(v)}$, no further operations are to be carried out on the numerator and denominator, viz. no cancelling by eventual common divisors which are functions in w (with the exception of rational integers of the coefficients , no rationalization of the denominator or multiplication of the numerator and denominator by the same polynomial in w; the $a_i^{(v)}(w)$ will then, generally, have the form

$$(5.52) \quad \begin{cases} a_i^{(v)}(w) = \dfrac{m \sum_{j=0}^{s} h_{i,j}^{(v)} w^{s-j}}{q \sum_{j=0}^{t} k_{i,j}^{(v)} w^{t-j}} ; & (i=1,\ldots,n-1) \\ q,m,h_{i,j}^{(v)},k_{i,j}^{(v)} \text{ rational integer;} & (s,t,v=0,1,\ldots) \\ (q,m) = (h_{i,j}^{(v)},\ldots,h_{i,s}^{(v)}) = (k_{i,1}^{(v)},\ldots,k_{i,t}^{(v)}) = 1. \end{cases}$$

Then the components $b_i^{(v)}$ of $b^{(v)}$ are obtained by the inner-outer T-function

$$(5.53) \qquad b_i^{(v)} = \left[mq^{-1}[w]^{s-t}\right] \cdot \left[qm^{-1}[w]^{t-s}\, a_i^{(v)}[w]\right]$$

in symbols

$$(5.54) \qquad f(a^{(k)}) = b^{(k)} = \left[a^{(k)}[w]\right].$$

The reader should note that, both the inner and the outer T-functions are special cases of the inner-outer T-function.

We shall illustrate the inner-outer T-function with a few examples. Let w be an algebraic irrationality and denote $[w] = D$, let be

(i) $a_i^{(v)}(w) = mq^{-1}w^i;$ $\quad m,q \in N;$ $\quad q|D$

then

$$b_i^{(v)} = \left[mq^{-1}D^i\right] \cdot \left[qm^{-1}D^{-i}mq^{-1}D^i\right]$$

$$= \left[mq^{-1}D^i\right] = mq^{-1}D^i.$$

(ii) $$a_i^{(v)}(w) = \frac{qw^s}{m(w^t + Dw^{t-1} + \cdots + D^t)}; \qquad (0 \leqslant s < t = 1,\ldots,n-1)$$

$$b_i^{(v)} = \left[qm^{-1}D^{s-t}\right]\left[mq^{-1}D^{t-s} \cdot \frac{q\,D^s}{m(t+1)\;D^t}\right]$$

$$= \left[qm^{-1}D^{s-t}\right]\left[\frac{1}{t+1}\right] = 0,$$

since $t \geqslant 1$ and $\left[\frac{1}{t+1}\right] = 0$.

(iii) $$a_i^{(v)}(w) = \frac{q(w^{s+1}+Dw^s+\cdots+D^s w + D^{s+1})}{m(w^s + Dw^{s-1}+\cdots+D^{s-1}w + D^s)}. \qquad (s=1,\ldots,n-2)$$

$$b_i^{(v)}(w) = \left[qm^{-1}D^{s+1-s}\right]\left[mq^{-1}D^{s-(s+1)}\;\frac{q(s+2)D^{s+1}}{m(s+1)D^s}\right]$$

$$= \left[qm^{-1}\,D\right]\left[\frac{s+2}{s+1}\right] = \left[qm^{-1}D\right],$$

since $\left[\frac{s+2}{s+1}\right] = 1$ for $s \geqslant 1$.

In order to state the main results of this chapter we shall again use a special notation for a matrix.

Definition IX. An $(n-1)$ by $(n-1)$ matrix of the form

$$\begin{pmatrix} s_1 & s_2 & \cdots & s_{n-1} \\ 0 & \ddots & & t_2 \\ & \ddots & \ddots & \vdots \\ & & 0 & t_{n-1} \end{pmatrix} \tag{5.55}$$

where s_i $(i=1,\ldots,n-1)$, t_i $(i=2,\ldots,n-1)$ are natural numbers, is called a B-fugue; the first row of a B-fugue is called the row-genus, its last column - the column-genus.

THEOREM 14. Let denote

$$(5.56) \qquad w = \sqrt[n]{D^n + d}; \quad d,D,n \in N; \quad d|D; \quad n \geqslant 3.$$

Then the JPA with the inner-outer T-function of the vector

$$(5.57) \qquad a^{(0)} = w^{n-1}, w^{n-2}, \ldots, w^2, w+D)$$

is purely periodic; the length of the primitive period is $n(n-1)$ (n B-fugues) for $d \neq 1$ and $n-1$ (one B-fugue) for $d = 1$. In case $d \neq 1$, the row-genus and the column-genus of the first B-fugue have the form

$$(5.58) \qquad \left\{ \begin{array}{lllll} D^{n-1}, & D^{n-2}, & \ldots, & D^2, & 2D \\ & & & & d^{-1}D \\ & & & & D \\ & & & & \vdots \\ & & & & D \end{array} \right. .$$

In the k+1-st B-fugue the row-genus and the column-genus have the form

$$(5.59) \qquad \left\{ \begin{array}{lllllll} d^{-1}D^{n-1}, & \ldots, & d^{-1}D^{n-k}, & D^{n-1-k}, & \ldots, & D^2, & 2D \\ & & & & & & D \\ & & & & & & \vdots \\ & & & & & & D \\ & & (k+2\text{-nd row}) & \ldots & \ldots & \ldots & d^{-1}D \\ & & & & & & D \\ & & & & & & \vdots \\ & & & & & & D \\ & (k=1,2,\ldots,n-3). & & & & & \end{array} \right.$$

In the n-1-st B-fugue the row-genus and the column-genus have the form

(5.60)
$$\begin{cases} d^{-1}D^{n-1}, & d^{-1}D^{n-2}, & \ldots, & d^{-1}D^{2}, & 2D \\ & & & & D \\ & & & & \vdots \\ & & & & D \end{cases}$$

In the n-th B-fugue the row-genus and the column-genus have the form

(5.61)
$$\begin{cases} d^{-1}D^{n-1}, & d^{-1}D^{n-2}, & \ldots, & d^{-1}D^{2}, & 2Dd^{-1} \\ & & & & D \\ & & & & \vdots \\ & & & & D \end{cases}$$

In case $d = 1$, the row-genus and the column-genus of the only B-fugue have the form

$$\begin{cases} D^{n-1}, & D^{n-2}, & \ldots, & D^{2}, & 2D \\ & & & & D \\ & & & & \vdots \\ & & & & D\ . \end{cases}$$

We leave the proof of Theorem 14 to the reader; it is obtained by induction and is quite lengthy. We shall illustrate Theorem 14 for the case

$$n = 3;\ w = \sqrt[3]{D^3 + d};\quad d|D.\quad a^{(0)} = (w^2, w+d).$$

$$b^{(0)} = (D^2,\ 2D);$$
$$b^{(1)} = (0,\ d^{-1}D);$$
$$b^{(2)} = (d^{-1}D^2,\ 2D);$$
$$b^{(3)} = (0,\ D);$$
$$b^{(4)} = (d^{-1}D^2,\ 2d^{-1}D);$$
$$b^{(5)} = (0,\ D).$$

We shall also state without proof

THEOREM 15. Let w be the (only) positive real root in the open interval $(D, D+1)$ of the polynomial

$$(5.62)\qquad \begin{aligned} &F(w) = w^n + (C-1)Dw^{n-1} - CD^n - d,\\ &C,D,d,n \in N;\quad d|D;\quad n \geqslant 3. \end{aligned}$$

Then the JPA with the inner-outer T-function of the vector

$$a^{(0)} = (w^{n-1}, w^{n-2}, \ldots, w^2, w+D)$$

is purely periodic, and the length of the primitive period is $n(n-1)$ for $d \neq 1$ and $n-1$ for $d = 1$. The period has the form of the fugues as in (5.58), (5.59), (5.60), (5.61), where in the second element of each column-genus D is to be replaced by CD.

Chapter 6.

UNITS IN ALGEBRAIC NUMBER FIELDS

§1. The Characteristic Equation of a Periodic JPA

In this paragraph we shall define and construct the characteristic equation associated with a periodic JPA, and prove the basic theorem about units of an algebraic field obtainable from such a periodic JPA. The characteristic equation will reveal one of the main fundamentals of periodic JPA, namely the fact that if the JPA, associated with any T-function, of a vector $a^{(0)} \in E_{n-1}$ becomes periodic, then the components of $a^{(0)}$ are all algebraic numbers of degree $\leq n$.

The basic tool of these investigations is formula (1.14), namely

$$a_i^{(0)} = \frac{A_i^{(v)} + a_1^{(v)}A_i^{(v+1)} + \cdots + a_{n-1}^{(v)}A_i^{(v+n-1)}}{A_0^{(v)} + a_1^{(v)}A_0^{(v+1)} + \cdots + a_{n-1}^{(v)}A_0^{(v+n-1)}}. \quad (i=1,\ldots,n-1;\ v=0,1,\ldots)$$

Since the $a_i^{(0)}$ are all well defined, the denominators of $a_i^{(0)}$ are different from zero. We shall first presume that the JPA, associated with any T-function, is purely periodic, and that the length of the primitive period is m. At a later stage we shall return to periodic JPA's with a pre-period. Writing m for v and denoting the numerator by y, we obtain from the above formula, recalling that, in view of pure periodicity

$$a_i^{(m)} = a_i^{(0)}, \qquad (i=1,\ldots,n-1) \tag{6.1}$$

$$\begin{cases} A_0^{(m)} + a_1^{(0)}A_0^{(m+1)} + a_2^{(0)}A_0^{(m+2)} + \cdots + a_{n-1}^{(0)}A_0^{(m+n-1)} = y, \\ A_1^{(m)} + a_1^{(0)}A_1^{(m+1)} + a_2^{(0)}A_1^{(m+2)} + \cdots + a_{n-1}^{(0)}A_1^{(m+n-1)} = a_1^{(0)}y, \\ A_2^{(m)} + a_1^{(0)}A_2^{(m+1)} + a_2^{(0)}A_2^{(m+2)} + \cdots + a_{n-1}^{(0)}A_2^{(m+n-1)} = a_2^{(0)}y, \\ \text{- -} \\ A_{n-1}^{(m)} + a_1^{(0)}A_{n-1}^{(m+1)} + a_2^{(0)}A_{n-1}^{(m+2)} + \cdots + a_{n-1}^{(0)}A_{n-1}^{(m+n-1)} = a_{n-1}^{(0)}y. \end{cases} \tag{6.2}$$

Multiplying $A_0^{(m)}, A_1^{(m)}, \ldots, A_{n-1}^{(m)}$, y by the factor $a_0^{(0)} = 1$, (6.2) represents a system of n homogeneous linear equations in the n variables $a_0^{(0)}, a_1^{(0)}, \ldots, a_{n-1}^{(0)}$, and since not all of the variables (actually none of them) vanish, the determinant of the system must vanish, and we obtain

$$(6.3) \qquad \begin{vmatrix} A_0^{(m)}-y, & A_0^{(m+1)}, & A_0^{(m+2)}, & \ldots, & A_0^{(m+n-1)} \\ A_1^{(m)}, & A_1^{(m+1)}-y, & A_1^{(m+2)}, & \ldots, & A_1^{(m+n-1)} \\ A_2^{(m)}, & A_2^{(m+1)}, & A_2^{(m+2)}, & \ldots, & A_2^{(m+n-1)} \\ - & - & - & - & - \\ A_{n-1}^{(m)}, & A_{n-1}^{(m+1)}, & A_{n-1}^{(m+2)}, & \ldots, & A_{n-1}^{(m+n-2)}, \; A_{n-1}^{(m+n-1)}-y \end{vmatrix} = 0.$$

Equation (6.3) is called the characteristic equation of the JPA in question. O. Perron has proved in [22] that the characteristic **equation is irreducible in the domain of rational functions (with rational coefficients) of the $A_i^{(m+v)}$ $(i,v = 0,\ldots,n-1)$. We can now obtain the $a_i^{(0)}$ from the system of the n-1 last linear equations of (6.2) ; the determinant of this system has the form**

$$(6.4) \qquad \begin{vmatrix} A_1^{(m+1)}-y, & A_1^{(m+2)}, & \ldots, & A_1^{(m+n-1)} \\ A_2^{(m+1)}, & A_2^{(m+2)}-y, & \ldots, & A_2^{(m+n-1)} \\ - & - & - & - \\ A_{n-1}^{(m+1)}, & A_{n-1}^{(m+2)}, & \ldots, & A_{n-1}^{(m+n-1)}-y \end{vmatrix} .$$

(6.4) is a polynomial in y of degree n-1, and since the characteristic equation is irreducible in the domain of rational functions of the $A_i^{(m+v)}$, (6.4) cannot vanish identically. Thus the $a_i^{(0)}$ $(i=1,\ldots,n-1)$ are expressed as polynomials in y and are therefore algebraic numbers of degree $\leq n$, since y is one.

Now let the JPA with any T-function be periodic with length ℓ of the pre-period and length m of the period; then

$$a_i^{(0)} = \frac{A_i^{(\ell)} + a_1^{(\ell)} A_i^{(\ell+1)} + \cdots + a_{n-1}^{(\ell)} A_i^{(\ell+m-1)}}{A_0^{(\ell)} + a_1^{(\ell)} A_0^{(\ell+1)} + \cdots + a_{n-1}^{(\ell)} A_0^{(\ell+m-1)}}, \quad (i=1,\ldots,n-1)$$

and since the JPA with the same T-function of the vector $(a_1^{(\ell)}, a_2^{(\ell)}, \ldots, a_{n-1}^{(\ell)})$ is purely periodic, the $a_i^{(\ell)}$ are algebraic numbers of degree $\leqslant n$, and so are therefore the $a_i^{(0)}$. We have proved

THEOREM 16. If the JPA, associated with some T-function (note that the $b^{(v)}$ have rational components) of the vector $a^{(0)} = (a_1^{(0)}, \ldots, a_{n-1}^{(0)})$ is periodic, then the components $a_i^{(0)}$ $(i=1,\ldots,n-1)$ of $a^{(0)}$ are all algebraic numbers of degree $\leqslant n$.

As we have pointed out repeatedly, the question whether the JPA with the outer T-function (or any other T-function) of any vector $a^{(0)}$ with algebraic components would always become periodic is still challengingly open.

If the primitive period is effectively given by the vectors $b^{(0)}, b^{(1)}, \ldots, b^{(m)}$, then the numbers $A_i^{(m+v)}$ $(i,v=0,\ldots,n-1)$ can be calculated, and hence y and the $a_i^{(0)}$ can be calculated in principle. But the $a_i^{(0)}$ cannot be, generally, stated explicitly, since y cannot. One point which has so far been completely overlooked must yet be clearly emphasized: an effectively given periodic JPA (with or without pre-period) and the vector $a^{(0)}$ derived from the characteristic equation reveal nothing about the T-function associated with the respective JPA; the T-function can only be guessed by testing the JPA. We shall illustrate this situation with a special case. Let the JPA of a vector $a^{(0)}$ be purely periodic, and let the primitive period consist of one vector

$$b^{(0)} = (k_1, k_2, \ldots, k_{n-1}) \in E_{n-1}. \tag{6.5}$$

Since here m = 1, we obtain from (6.3)

$$(6.6)\qquad \begin{vmatrix} A_0^{(1)}-y, & A_0^{(2)}, & & \ldots, & A_0^{(n-1)}, & A_0^{(n)} \\ A_1^{(1)}, & A_1^{(2)}-y, & & \ldots, & A_1^{(n-1)}, & A_1^{(n)} \\ A_2^{(1)}, & A_2^{(2)}, & A_2^{(3)}-y, & \ldots, & A_2^{(n-1)}, & A_2^{(n)} \\ - - - & - - - & - - - & - - - & - - - & - - - \\ A_{n-1}^{(1)}, & A_{n-1}^{(2)}, & & \ldots, & A_{n-1}^{(n-1)}, & A_{n-1}^{(n)}-y \end{vmatrix} = 0.$$

The reader will easily verify the calculation

$$(6.7)\qquad A_i^{(n)} = k_i. \qquad (i=0,\ldots,n-1)$$

In view of (6.7), equation (6.6) becomes

$$(6.8)\qquad \begin{vmatrix} -y, & 0, & 0, & & \ldots, & 0, & 1 \\ 1, & -y, & 0, & & \ldots, & 0, & k_1 \\ 0, & 1, & -y, & 0, & \ldots, & 0, & k_2 \\ - - - & - - - & - - - & - - - & - - - & - - - & - - - \\ 0, & 0, & \ldots, & & & 0,\ 1, & k_{n-1}y \end{vmatrix} = 0.$$

The determinant of (6.8) is well known; in expanded form it becomes

$$(6.9)\qquad y^n - k_{n-1}y^{n-1} - k_{n-2}y^{n-2} - \cdots - k_2y^2 - k_1y - 1 = 0.$$

We shall now specify the vector $b^{(0)}$ so that

$$(6.10)\qquad k_{n-i} = \binom{n}{i} D^{n-i}. \qquad (i=1,\ldots,n-1)$$

From (6.9) we obtain, in virtue of (6.10)

$$y^n - \left(\binom{n}{1}(Dy)^{n-1} + \binom{n}{2}(Dy)^{n-2} + \cdots + \binom{n}{n-2}(Dy)^2 + \binom{n}{n-1}Dy + 1\right) = 0,$$

$$y^n - \left[(Dy+1)^n - D^n y^n\right] = 0,$$

$$(D^n+1)y^n - (Dy+1)^n = 0, \qquad \sqrt[n]{D^n+1}\ y = Dy + 1$$

$$(6.11)\qquad y = \frac{1}{w-D}; \qquad w = \sqrt[n]{D^n+1}.$$

The equations (6.2) take the form

$$(6.12)\quad \begin{cases} a_{n-1}^{(0)} = y, \\ 1 + \binom{n}{1} D a_{n-1}^{(0)} = a_1^{(0)} y, \\ a_1^{(0)} + \binom{n}{2} D^2 a_{n-1}^{(0)} = a_2^{(0)} y, \\ a_2^{(0)} + \binom{n}{3} D^3 a_{n-1}^{(0)} = a_3^{(0)} y, \\ - - - - - - - - - - - - - - \\ a_{n-3}^{(0)} + \binom{n}{n-2} D^{n-2} a_{n-1}^{(0)} = a_{n-2}^{(0)} y. \end{cases}$$

From the first of the equations (6.12) we obtain

$$a_{n-1}^{(0)} = \frac{1}{w-D} = w^{n-1} + Dw^{n-2} + \cdots + D^{n-2}w + D^{n-1},$$

from the second

$$a_1^{(0)} = \frac{1+\binom{n}{1}Dy}{y} = \frac{1}{y} + \binom{n}{1}D = w-D + \binom{n}{1}D = w + \binom{n-1}{1}D;$$

from the third

$$a_2^{(0)} = \frac{a^{(0)} + \binom{n}{2}D^2 y}{y} = (w-D)\left(w + \binom{n-1}{1}D\right) + \binom{n}{2}D^2$$

$$= w^2 + \binom{n-2}{1}Dw + \left(\binom{n}{2} - \binom{n-1}{1}\right)D^2$$

$$= w^2 + \binom{n-2}{1}Dw + \binom{n-1}{2}D^2.$$

We thus have proved for $s = 1,2$

$$(6.13)\qquad a_s^{(0)} = \sum_{i=0}^{s} \binom{n-1-s+i}{i} w^{s-i} D^i.$$

Suppose this formula is correct for $s = t \leqslant n-2$; we then obtain from (6.12)

$$a_{t+1}^{(0)} = \frac{a_t^{(0)} + \binom{n}{t+1}D^{t+1}y}{y}$$

$$= (w-D) \sum_{i=0}^{t} \binom{n-1-t+i}{i} w^{t-i} D^i + \binom{n}{t+1}D^{t+1}$$

$$= \sum_{i=0}^{t} \binom{n-1-t+i}{i} w^{t+1-i} D^i - \sum_{i=0}^{t-1} \binom{n-1-t+i}{i} w^{t-i} D^{i+1}$$

$$- \binom{n-1}{t} D^{t+1} + \binom{n}{t+1} D^{t+1}$$

$$= w^{t+1} + \sum_{i=1}^{t} \binom{n-1-t+i}{i} w^{t+1-i} D^i - \sum_{i=1}^{t} \binom{n-2-t+i}{i} w^{t+1-i} D^i$$

$$+ \binom{n-1}{t+1} D^{t+1}$$

$$= w^{t+1} + \sum_{i=1}^{t} \binom{n-1-(t+1)+i}{i} w^{t+1-i} D^i + \binom{n-1}{t+1} D^{t+1}$$

$$= \sum_{i=0}^{t+1} \binom{n-1-(t+1)+i}{i} w^{t+1-i} D^i ,$$

so that formula (6.13) is correct for $s = 1,\dots,n-1$. So, $a^{(0)} = (\dots, \sum_{i=0}^{s} \binom{n-1-s+i}{i} w^{s-i} D^i, \dots)$ is the vector in E_{n-1}, where JPA is purely periodic with the primitive period $b^{(0)} = (\binom{n}{1} D, \binom{n}{2} D^2, \dots, \binom{n}{n-1} D^{n-1})$. But what is the T-function of this JPA; as long as $D \geqslant n-2$ (since $d = 1$) and D is a natural number, it is the outer T-function; but D can be any real number, and then another T-function must be associated with this JPA.

§ 2. Units and Periodicity of the JPA

From the determinantal form of the characteristic equation (6.3) three basic facts about its coefficients are immediately obvious:

(i) its leading coefficient is $(-1)^n$;

(ii) its free term is the determinant

$$\begin{vmatrix} A_0^{(m)}, & A_0^{(m+1)}, & \dots, & A_0^{(m+n-1)} \\ A_1^{(m)}, & A_1^{(m+1)}, & \dots, & A_1^{(m+n-1)} \\ - & - & - & - \\ A_{n-1}^{(m)}, & A_{n-1}^{(m+1)}, & \dots, & A_{n-1}^{(m+n-1)} \end{vmatrix} = (-1)^{m(n-1)};$$

(iii) its coefficients are rational integers, if the components of the vectors

$$b^{(0)},\ b^{(1)},\ \ldots,\ b^{(m-1)}$$

are rational integers; for then also the numbers $A_i^{(m+v)}$ $(i,v=0,\ldots,1)$ are rational integers.

We can therefore write the characteristic equation in the form

$$(6.14)\qquad y^n + c_1y^{n-1} + c_2y^{n-2} + \cdots + c_{n-1}y + (-1)^{m(n-1)+n} = 0.$$

If the c_i $(i=1,\ldots,n-1)$ are rational integers, then we can read off from (6.14) the most significant result that y is a unit in $Q(y)$ (the field generated by adjunction of the irrational y to the field of rationals). Since $a_1^{(0)}, a_2^{(0)}, \ldots, a_{n-1}^{(0)}$ are all polynomials in y, the field $Q(a_1^{(0)}, \ldots, a_{n-1}^{(0)})$ is identical with $Q(y)$. We shall denote this field by $Q(w)$, $a_i^{(0)} = a_i^{(0)}(w)$, where w is a polynomial **of degree n in y** . Here is an example. From the characteristic equation (6.9), we obtain from (6.11)

$$w - D = \frac{1}{y}$$

$$w = D + y^{n-1} - k_1y^{n-2} - k_2y^{n-3} - \cdots - k_{n-1}.$$

We recall at this moment, that the JPA in question is purely periodic. We further obtain from (6.2)

$$(6.15)\qquad y = A_0^{(m)} + a_1^{(0)}A_0^{(m+1)} + \cdots + a_{n-1}^{(0)}A_0^{(m+n-1)},$$

and, returning to formula (1.16)

$$y = \sum_{i=0}^{n-1} a_i^{(0)}A_0^{(m+i)} = \prod_{i=1}^{m} a_{n-1}^{(i)}.$$

But since the JPA is purely periodic, with length of period m, we have $a_{n-1}^{(m)} = a_{n-1}^{(0)}$, so that

$$(6.16)\qquad y = \sum_{i=0}^{n-1} a_i^{(0)}A_0^{(m+i)} = \prod_{i=0}^{m-1} a_{n-1}^{(i)}.$$

If the JPA of $a^{(0)}$ is not purely periodic, then (if again ℓ denotes the length of the pre-period and m denotes the length of the period of this periodic JPA), then the JPA of $a^{(\ell)}$ is purely periodic, and

$$(6.17) \qquad e = \prod_{i=\ell}^{\ell+m-1} a_{n-1}^{(i)}$$

is a unit in the field $Q(w)$, where $a^{(0)} = a^{(0)}(w)$. We have thus obtained the main results of periodic JPA connected with units of the respective algebraic number field which we state in

THEOREM 17. Let $a^{(0)} = a^{(0)}(w) \in E_{n-1}$ be a vector with algebraic components of degree $\leqslant n$. If the JPA of $a^{(0)}$ is periodic and the associated T-function is such that the components of the vectors $b^{(v)}$ $(v=0,1,\dots)$ are all rational integers, then the product of the last components of the vectors $a^{(\ell)}, a^{(\ell+1)}, \dots, a^{(\ell+m-1)}$ (ℓ - length of pre-period, m - length of period of the JPA of $a^{(0)}$) is a unit of the field $Q(w)$.

In [3,a] Helmut Hasse and the author gave a different proof of this theorem; another proof was again given by the author in [2,d]. But in these previous proofs the authors operated with a JPA associated with the outer T-function $f(a^{(k)}) = \left[a^{(k)}\right]$. Theorem 17 has the advantage over the previous analogous results, that all that is demanded here is the integrality of the components of the $b^{(v)}$ $(v=0,1,\dots)$.

For a periodic JPA with a pre-period we shall state a formula similar to that of (6.17). We obtain from (1.16)

$$\prod_{i=1}^{\ell+m-1} a_{n-1}^{(i)} = \sum_{j=0}^{n-1} a_j^{(\ell+m-1)} A_0^{(\ell+m-1+j)}$$

$$\prod_{i=1}^{\ell-1} a_{n-1}^{(i)} = \sum_{j=0}^{n-1} a_j^{(\ell-1)} A_0^{(\ell-1+j)}$$

$$(6.18)\qquad e = \prod_{i=\ell}^{\ell+m-1} a_{n-1}^{(i)} = \frac{\sum_{j=0}^{n-1} a_j^{(\ell+m-1)} A_0^{(\ell+m-1+j)}}{\sum_{j=0}^{n-1} a_j^{(\ell-1)} A_0^{(\ell-1+j)}}. \qquad (\ell \geqslant 1)$$

Let again the JPA of $a^{(0)}$ be purely periodic. We shall give a formula for the powers of e. We obtain from (1.16)

$$\prod_{i=1}^{sm} a_{n-1}^{(i)} = \sum_{j=0}^{n-1} a_j^{(sm)} A_0^{(sm+j)}, \qquad s \geqslant 1.$$

But

$$\prod_{i=1}^{sm} a_{n-1}^{(i)} = \prod_{t=0}^{s-1} \prod_{i=1}^{m} a_{n-1}^{(tm+i)},$$

and since $a_{n-1}^{(tm+i)} = a_{n-1}^{(i)}$, $a_{n-1}^{(m)} = a_{n-1}^{(0)}$

$$\prod_{i=1}^{sm} a_{n-1}^{(i)} = \prod_{t=0}^{s-1} \prod_{i=1}^{m} a_{n-1}^{(i)} = \prod_{t=0}^{s-1} \prod_{i=0}^{m-1} a_{n-1}^{(i)} = \prod_{t=0}^{s-1} e = e^s,$$

so that, **by the property of the m-th powers of eigenvalues ,**

$$(6.19)\qquad e^s = \sum_{j=0}^{n-1} a_j^{(0)} A_0^{(sm+j)}. \qquad (s=1,2,\ldots)$$

From (6.19) we obtain, on basis of (1.15)

$$e^{-s} = \left(\sum_{j=0}^{n-1} a_j^{(0)} A_0^{(sm+j)} \right)^{-1} = \left(\sum_{j=0}^{n-1} a_j^{(sm)} A_0^{(sm+j)} \right)^{-1}$$

$$= \begin{vmatrix} 1 & A_0^{(sm+1)} & \ldots & A_0^{(sm+n-1)} \\ a_1^{(0)} & A_1^{(sm+1)} & \ldots & A_0^{(sm+n-1)} \\ \vdots & \vdots & & \vdots \\ a_{n-1}^{(0)} & A_{n-1}^{(sm+1)} & \ldots & A_{n-1}^{(sm+n-1)} \end{vmatrix}$$

$$(6.20)\qquad e^{-s} = \begin{vmatrix} 1 & A_0^{(sm+1)} & \cdots & A_0^{(sm+n-1)} \\ a_1^{(0)} & A_1^{(sm+1)} & \cdots & A_1^{(sm+n-1)} \\ \vdots & \vdots & & \vdots \\ a_{n-1}^{(0)} & A_{n-1}^{(sm+1)} & \cdots & A_{n-1}^{(sm+n-1)} \end{vmatrix}$$

$(s=1,2,\ldots)$

We also see from (6.19), that $e^s > 1$ for $s \geqslant 1$.

§ 3. Explicit Units of Algebraic Number Fields

In this section we shall state explicitly units of algebraic number fields based on periodicity of the JPA of various vectors $a^{(0)}$. It should be noted that so far we have made no comments whatsoever, as to whether or not the unit gained from a periodic JPA **belongs to a system of fundamental units. We discuss this subject at the end** of this chapter. Throughout this section we shall operate with purely periodic JPA's only; consequently we shall calculate a unit of the respective algebraic number field from (6.16), viz.

$$e = \prod_{i=0}^{m-1} a_{n-1}^{(i)}.$$

Since $a_{n-1}^{(k+1)} = \frac{1}{a_1^{(k)} - b_1^{(k)}}$, $(k=0,1,\ldots)$ we shall use, instead of the above formula

$$(6.21)\qquad e = a_{n-1}^{(0)} \prod_{k=0}^{m-2} \left(a_1^{(k)} - b_1^{(k)}\right)^{-1}.$$

We shall first investigate a most general case. Let $F(x)$ be an n-th degree polynomial of the form

$$(6.22)\qquad \begin{aligned} &F(x) = x^n + k_1 x^{n-1} + \cdots + k_{n-1} x - d; \qquad (n \geqslant 2) \\ &k_i \ (i=1,\ldots,n-1), \quad d \in I; \quad d \mid k_i; \quad d \neq 0. \end{aligned}$$

Let w be any irrational real root of $F(x)$, viz.

(6.23) $$w^n + k_1 w^{n-1} + \cdots + k_{n-1} w - d = 0$$

and choose the vector $a^{(0)} \in E_{n-1}$

(6.24) $$a^{(0)} = (w+k_1,\ w^2+k_1w+k_2, \ldots, w^{n-1}+k_1w^{n-2}+\cdots+k_{n-2}w+k_{n-1}).$$

Then, as we know, the JPA with the T-function

(6.25) $$b_i^{(v)} = a_i^{(v)}(w)_{(w=0)} \qquad (i=1,\ldots,n-1;\ v=0,1,\ldots)$$

is purely periodic and the length of the period is n for $d \neq 1$ and n for $d = 1$. The period has the form (3.11). We obtain from (6.23), (6.24)

$$a_{n-1}^{(0)} = w^{n-1}+k_1w^{n-2}+\cdots+ k_{n-2}w + k_1 = \frac{d}{w};$$

the reader will verify easily that

$$a_1^{(v)} - b_1^{(v)} = w, \qquad (v=0,\ldots,n-2)$$

so that, in virtue of (6.21),

$$e = \frac{d}{w} \prod_{v=0}^{n-2} w^{-1},$$

(6.26) $$e = \frac{d}{w^n}, \quad e^{-1} = \frac{w^n}{d}.$$

We have thus obtained

THEOREM 18. Let w be a positive root of the polynomial (6.22). Then $e^{-1} = d^{-1}w^n$ is a unit in $Q(w)$.

We shall now specify the coefficients k_i $(k=1,\ldots,n-1)$ of the polynomial (6.22); let be, as before,

(6.27) $$k_i = \binom{n}{i} D^i; \quad D \in N, \quad d \mid D; \quad n \geqslant 2;$$

then

$$F(w) = w^n + \binom{n}{1} D w^{n-1} + \cdots + \binom{n}{n-1} D^{n-1} w - d$$
$$= (w + D)^n - D^n - d = 0,$$

(6.28) $$w = \sqrt[n]{D^n + d} - D = \alpha - D.$$

On the basis of (6.26), (6.27), (6.28) we now obtain the important

Corollary 1 (to Theorem 18). Let be

$$D, k, n \text{ natural numbers}; \quad (n \geqslant 2)$$
(6.29)
$$\alpha = \sqrt[n]{D^n \pm k};$$

then

(6.30) $$\varepsilon = \frac{(\alpha - D)^n}{k} = \frac{(\alpha - D)^n}{\alpha^n - D^n}$$

is a unit in $Q(\alpha)$ for the following values of k:

(6.31)
(i) $k = d;\quad d \mid D$
(ii) $k = pd,\ d \mid D;\quad n = p^v\ (v=1,2,\ldots,);$ p a prime.

The case $\alpha = \sqrt[3]{D^3 + 3D}$ is obtained from (6.31), (ii) for $p = n = 3;\ d = D$, and the unit of $Q(\alpha)$ has the form

$$\varepsilon = \frac{(\alpha - D)^3}{3D} = \frac{\alpha^3 - D^3 - 3D\alpha^2 + 3D^2\alpha}{3D}$$

$$\varepsilon = -\alpha^2 + 3D\alpha + 1.$$

Now by Theorem 6, the JPA with the outer T-function of the vector $a^{(0)} = (\alpha, \alpha^2)$, $\alpha = \sqrt[3]{D^3 + D}$ is periodic with length 4 of both the pre-period and the period. We leave it as an exercise to the reader to prove that the product $a^{(4)}_{n-1} a^{(5)}_{n-1} a^{(6)}_{n-1} a^{(7)}_{n-1}$ of this JPA supplies the same unit as the above. Formula (6.31) does not include the case $\alpha = \sqrt[3]{D^3 + 6D}$, $D = 2k$, $k = 2,3,\ldots$. By Theorem 7, the JPA with the outer T-function of $a^{(0)} = (\alpha, \alpha^2)$ is periodic with length of pre-period 4 and length of period 8. It would be most cumbersome to calculate the unit of the field $Q(\sqrt[3]{D^3 + 6D})$ from the second components of the vectors $a^{(4)}, \ldots, a^{(11)}$. We shall achieve this by a different method. We shall take again refuge to the inner T-function and obtain

$$a^{(0)} = (\alpha + 2D,\ \alpha^2 + D\alpha + D^2);$$

$$b^{(0)} = (3D,\ 3D^2);$$

$$a^{(1)} = (\alpha + 2D,\ \frac{\alpha^2 + D\alpha + D^2}{6D});$$

$$b^{(1)} = (3D,\ \frac{3D}{2});$$

$$a^{(2)} = \frac{\alpha + 2D}{6D},\ \frac{\alpha^2 + D\alpha + D^2}{6D};$$

$$b^{(2)} = (\frac{1}{2},\ \frac{3D}{2});$$

$$a^{(3)} = (\alpha + 2D,\ \alpha^2 + D\alpha + D^2) = a^{(0)}.$$

So, the JPA with the outer T-function of $a^{(0)} = (\alpha + 2D, \alpha^2 + D\alpha + D^2)$ is purely periodic and the length of the period is 3. Since $2|D$, the components of $b^{(0)}$, $b^{(1)}$, $b^{(2)}$ are all rational integers, with the exception of $b_1^{(2)}$. **This does not** contradict Theorem 17; **for** the conditions stated herein, namely that the $b_i^{(v)}$ $(i=1,\dots,n-1; v=0,\dots,m-1)$ be integers, are only sufficient conditions; a necessary and sufficient condition for

$$y = \prod_{i=0}^{m-1} a_{n-1}^{(i)} = \sum_{i=0}^{n-1} a_i^{(0)} A_0^{(m+1)}$$

to be a unit is that the coefficients of the characteristic equation (6.14) be integers, or that y^{-1} be an algebraic integer and the norm of y^{-1}, $N(y^{-1}) = \pm 1$. Let be $y^{-1} = \frac{(w-D)^n}{k}$, $w = \sqrt[n]{D^n + k}$. In the next chapter it will be proved, that then

$$N(w-D) = \pm k,$$

so that

$$N(y^{-1}) = \frac{\pm k^n}{k^n} = \pm 1.$$

Thus, if $y^{-1} = \frac{(w-D)^n}{k}$, $w = \sqrt[n]{D^n + k}$, y^{-1} is a unit if it is an algebraic integer.

$$a_{n-1}^{(0)} = \alpha^2 + D\alpha + D^2 = \frac{6D}{\alpha - D};$$

$$a_{n-1}^{(1)} = \frac{\alpha^2 + D\alpha + D^2}{6D} = \frac{1}{\alpha - D};$$

$$a_{n-2}^{(0)} = \frac{\alpha^2 + D\alpha + D^2}{6D} = \frac{1}{\alpha - D};$$

$$y = a_{n-1}^{(0)}a_{n-1}^{(1)}a_{n-1}^{(2)} = \frac{6D}{(\alpha - D)^3};$$

$$y^{-1} = \frac{(\alpha - D)^3}{6D} = \frac{6D - 3D\alpha^2 + 3D^2\alpha}{6D};$$

$$y^{-1} = -\frac{\alpha^2}{2} + \frac{D}{2}\alpha + 1.$$

To show that y^{-1} is an algebraic integer, we need only to prove that $\frac{\alpha^2}{2}$ is an algebraic integer, since α is one and $2|D$. Let be

$$\frac{\alpha^2}{2} = x,\ \alpha^2 = 2x,\ \alpha^6 = 8x^3,\quad (D^3 + 5D)^2 = 8x^3,$$

$$D^6 + 12D^4 + 36D^2 = 64k^6 + 192k^4 + 144k^2 = 8x^3,$$

$$x^3 = 8k^6 + 24k^4 + 18k^2,$$

so x^3, and therefore x are algebraic integers, which proves that y^{-1} is an algebraic integer, and formula (6.30) is valid for the case $k = 6D$.

Indeed, this reasoning will enable us to find a few more infinite classes of a unit in algebraic number fields. We state these new results in

Corollary 2 (to Theorem 18). Let be D,k,n,α as in Corollary 1, then

$$\varepsilon = \frac{(\alpha - D)^n}{k} = \frac{(\alpha - D)^n}{\alpha^n - D^n}$$

is a unit in $Q(\alpha)$ for the following values of k

$$(6.32)\quad \begin{array}{l}\text{(i)}\quad k = d^rD \text{ or } d^r;\quad d|D;\quad r = 0,1,\ldots,n-1\\ \text{(ii)}\quad n = p^v\ (p\text{ prime, } v=1,2,\ldots);\ d = pd^rD \text{ or } pd^r;\ d|D.\end{array}$$

Proof. We shall verify (i). Following the ideas developed in in the case $\alpha = \sqrt[3]{D^3 + 6D}$ we have only to show that ε is an algebraic integer. We obtain

$$\frac{(\alpha - D)^n}{k} = \frac{\sum_{i=0}^{n} (-1)^i \binom{n}{i} \alpha^{n-i} D^i}{d^r D},$$

and we shall prove that

$$\frac{\alpha^{n-i} D^i}{d^r D}$$

is an algebraic integer for $i=0,1,\ldots,n$.

Let be

$$\frac{\alpha^{n-i} D^i}{d^r D} = x,$$

then, raising the equation to the n-th power

$$(\alpha^n)^{n-i} D^{in} = d^{rn} D^n x^n,$$

$$(D^n + d^r D)^{n-i} D^{in} = d^{rn} D^n x^n;$$

denoting $t = d^{-1}D$, we obtain,

$$d^{r(n-i)} D^{n-i} (D^{n-r} t^r + 1) D^{in} = d^{rn} D^n x^n,$$

$$d^{r(n-i)} D^{in-i} (D^{n-r} t^r + 1) = d^{rn} x^n,$$

$$d^{r(n-i)+in-i} t^{in-i} (D^{n-r} t^r + 1) = d^{rn} x^n,$$

and we have to prove

$$r(n-i) + in - i \geqslant rn$$

or

$$i(n - 1 - r) \geqslant 0,$$

which is obvious, since $i \geqslant 0$ and $n - 1 \geqslant r$. This verifies (i) of Corollary 2. To prove (ii), one has only to remember that

$p \mid \binom{p^v}{i}$, for $i = 1,\ldots,p^v - 1$. The case $\alpha = \sqrt[3]{D^3 + 6D}$ is included in (ii) for $p = 3$, $d \mid D$, $d = 2$, $D = 2k$.

§4. Sets of Independent Units in $Q(w)$

In the previous sections we have shown how one unit of an algebraic field can be obtained from a periodic JPA of a vector with components in this field. In this section we shall obtain a set of independent units for some specified infinite classes of algebraic fields.

Let be

(6.32) $$\alpha = \sqrt[n]{D^n + d}; \quad n = st.$$

Denote

(6.33) $$D^s = D'; \quad d = d'; \quad \alpha^s = w.$$

Then, from (6.32), (6.33)

(6.34) $$w = \sqrt[t]{D'^t + d'} .$$

On basis of Corollary 2 (to Theorem 18) the JPA with the inner T-function of $a^{(0)'} = (\ldots, \sum_{i=0}^{p} \binom{t-j-1+i}{i} w^{t-i} D'^i, \ldots)$ where the element in brackets is $a_j^{(0)'}$ $(j=1,\ldots,t-1)$ is purely periodic with length of period t, and

$$\varepsilon_t = \frac{(w - D')^t}{w^t - D'^t}$$

is a unit of $Q(w)$. But $Q(w) \subset Q(\alpha)$, so ε_t is a unit in $Q(\alpha)$. On basis of (6.33) ε_t takes the form

(6.35) $$\varepsilon_t = \frac{(\alpha^s - D^s)^t}{\alpha} .$$

But $\varepsilon = \frac{(\alpha - D)^n}{\alpha}$, so

$$\frac{\varepsilon}{\varepsilon_t} = \frac{(\alpha - D)^n}{(\alpha^s - D^s)^t} = \left(\frac{(\alpha - D)^s}{\alpha^s - D^s} \right)^t .$$

But $\frac{(\alpha - D)^s}{\alpha^s - D^s}$ is already in $Q(\alpha)$, therefore it is a unit in $Q(\alpha)$. We have thus obtained the important

THEOREM 19. Let be

$$\alpha = \sqrt[n]{D^n + d}; \quad n = st, \quad s > 1; \quad d|D.$$

Then $\tau(n) - 1$ units in $Q(\ )$ are given by the formula

(6.36)
$$\varepsilon^{(s)} = \frac{(\alpha - D)^s}{\alpha^s - D^s} .$$

Following Dirichlet's famous theorem about the maximal number of independent units in an algebraic number field, $Q(m^{1/4})$ has two fundamental units, if m is not the square of a rational number, $Q(m^{1/6})$ has three fundamental units if m is neither the square nor the cube of a rational. It is conjectured that in $Q(\alpha)$, $\alpha = (D^4 + d)^{1/4}$, the two units

$$\varepsilon = \frac{(D - \alpha)^4}{\alpha^4 - D^4} ,$$

$$\varepsilon^{(2)} = \frac{(D - \alpha)^2}{\alpha^2 - D^2}$$

and that in $Q(\alpha)$, $\alpha = (D^6 + d)^{1/6}$, the three units

$$\varepsilon = \frac{(\alpha - D)^6}{\alpha^6 - D^6} ,$$

$$\varepsilon^{(3)} = \frac{(\alpha - D)^3}{\alpha^3 - D^3} ,$$

$$\varepsilon^{(2)} = \frac{(\alpha - D)^2}{\alpha^2 - D^2} ,$$

form a system of independent units. These two cases (and, of course, the quadratic and the cubic case) are the only ones, where the number of fundamental units of the field equals the number of units obtainable from (6.36).

We shall now investigate an infinite class of algebraic number fields $Q(w)$ of degree $n \geqslant 2$ for which a maximal set of independent units is stated. These results were published in [3,b], and here the basic theorems of this paper are mostly given without proof. Our investigations are based on a third order P-polynomial described in

Definition X. A polynomial $F(x)$ of degree $n \geqslant 2$ is called a third order P-polynomial if it has the form

$$(6.37)\quad \begin{cases} F(x) = (x - D)(x - D_1)\dots(x - D_{n-1}) - d; \\ D, D_i, d \text{ rational integers; } D \equiv D_i \pmod d;\ (i=1,\dots,n-1); \\ d \geqslant 1;\ D = D_0 > D_1 > \cdots > D_{n-1}; \\ D_1 - D_2 \geqslant 2 \text{ or } D_0 - D_1 \geqslant 4 \quad \text{for } n = 3 \text{ and } d = 1; \\ D_1 - D_2 \geqslant 2 \text{ or } D_0 - D_1 \geqslant 3 \text{ or } D_2 - D_3 \geqslant 3 \\ \text{or } D_0 - D_1,\ D_2 - D_3 \geqslant 2 \text{ for } n = 4 \text{ and } d = 1. \end{cases}$$

The basic algebraic property of a third order P-polynomial is given by

Lemma 3. A third order P-polynomial has exactly n different real roots; of these lie:

One in the open interval $(D_0;\infty)$, more exactly in the open interval $(D_0;\ D_0 + 1)$;

two in each of the open intervals $(D_{2i},\ D_{2i-1})$, more exactly one in the open left half, and one in the open right half of these intervals with $2 \leqslant 2i \leqslant n-1$;

one in the open interval $(-\infty;D_{n-1})$ if n is even.

Rearranging $F(x)$ from (6.37) in powers of $x - D$, (where D is the greatest among the n roots of $F(x)$), the following formula is easily proved

$$(6.38)\quad \begin{cases} F(x) = -d + \sum_{s=0}^{n-1} k_s (x - D)^{n-s}, & (k_0 = 1) \\ k_s = \sum (D - D_{i_1})(D - D_{i_2}) \cdots (D - D_{i_s}) & \\ 1 \leqslant i_1 < i_2 < \cdots < i_s \leqslant n-1. & (s=1,\ldots,n-1) \end{cases}$$

On basis of (6.37) it is obvious that the k_s $(s=1,\ldots,n-1)$ are all positive numbers and that further $d | k_s$. It is now easily seen that the polynomial $F(x)$ in its form (6.38) is a first order P-polynomial, and we could now proceed along the lines of Theorem 3 to obtain a periodic JPA with the outer T-function. Yet we shall turn into a different direction because of reasons which will soon become obvious. We shall use B-fugues ((n-1) by (n-1)) matrices of the form

$$(6.39)\qquad A = \begin{pmatrix} 0 & \cdot & \cdot & \cdot & 0 & A_1 \\ & \cdot & & & & \vdots \\ & & \cdot & & & \\ & & & \cdot & 0 & A_{n-1} \end{pmatrix}$$

to state

THEOREM 20. Let $F(x)$ be a third order P-polynomial and w its greatest real root which is unique in the open interval $(D, D+1)$; let denote

$$(6.40)\qquad \begin{aligned} &P_{ii} = P_i = w - D_i; && (i=1,\ldots,n-1) \\ &P_{i,k} = P_i P_{i+1} \cdots P_k. && (i \leqslant i \leqslant k \leqslant n-1) \end{aligned}$$

Let $a^{(0)} = a^{(0)}(w)$ be a vector with the components

$$(6.41)\qquad \begin{aligned} &a_s^{(0)} = d^{-1}(w - D)P_{1,1}P_{2+s,n-1}; \qquad (s=1,\ldots,n-3) \\ &a_{n-2}^{(0)} = d^{-1}(w - D)P_{1,1}; \quad a_{n-1}^{(0)} = P_{1,1}. \end{aligned}$$

Then the JPA with the outer T-function of $a^{(0)}$ is purely periodic, and its primitive length is $m = n(n-1)$ for $d \neq 1$ and $m = n-1$ for $d = 1$. The period of length $n(n-1)$ consists of n fugues of the form (6.39). The column-genus of the first fugue has the form

$$(6.42)\qquad \begin{cases} D - D_1, \\ D - D_2, \\ \text{-------} \\ D - D_{n-1}; \end{cases}$$

the column-genus of the r + 1 - st fugue ($r = 1,\dots,n-1$) has the form

$$(6.43)\qquad \begin{cases} D - D_1, \\ D - D_2, \ \dots \\ \dots\dots\dots\dots \\ D - D_{r-1} \\ d^{-1}(D - D_r), \\ D - D_{r+1}, \\ \dots\dots\dots \\ D - D_{n-1}; \end{cases}$$

the period of length $m = n-1$ consists of one fugue whose column-genus has the form (6.42).

Example 9. Let be

$$F(x) = x^6 - 3x^5 - 5x^4 + 15x^3 + 4x^2 - 12x - 1$$

or

$$F(x) = (x-3)(x-2)(x-1)\ x\ (x+1)(x+2) - 1.$$

This is a third order P-polynomial with $D = 3$, $D_1 = 2$, $D_2 = 1$, $D_3 = 0$, $D_4 = -1$, $D_5 = -2$, $d = 1$

$$F(w) = 0, \quad 3 < w < 4.$$

The JPA with the outer t-function of the vector
$a^{(0)} = (a_1^{(0)}, a_2^{(0)}, a_3^{(0)}, a_4^{(0)}, a_5^{(0)})$,

$$a_1^{(0)} = (w-3)(w-2)w(w+1)(w+2) = w^5 - 2w^4 - 7w^3 + 8w^2 + 12w,$$

$$a_2^{(0)} = (w-3)(w-2)(w+1)(w+2) = w^4 - 2w^3 - 7w^2 + 8w + 12,$$

$$a_3^{(0)} = (w-3)(w-2)(w+2) = w^3 - 3w^2 - 4w + 12,$$

$$a_4^{(0)} = (w-3)(w-2) = w^2 - 5w + 6,$$

$$a_5^{(0)} = w - 2,$$

is purely periodic and the length of the primitive period is $m = 5$; this has the form

$$b^{(0)} = (0, 0, 0, 0, 1),$$
$$b^{(1)} = (0, 0, 0, 0, 2),$$
$$b^{(2)} = (0, 0, 0, 0, 3),$$
$$b^{(3)} = (0, 0, 0, 0, 4),$$
$$b^{(4)} = (0, 0, 0, 0, 5).$$

For additional investigations we need

Lemma 4. A third order P-polynomial with the property

(6.44) $$D_0 - D_i \geqslant 2d(n-1), \qquad (i=1,\ldots,n-1)$$

is irreducible in the field of rationals.

We shall further introduce the notation

(6.45) $$\begin{cases} R_{i,i} = w - D_{i,i}; \ D_{i,i} \in \{D_0,\ldots,D_{n-1}\} \\ R_{i,j} = R_{i,i}R_{i+1,i+1} \cdots R_{j,j}; \\ R_{i,i} \neq R_{j,j} \quad \text{if} \quad i \neq j. \end{cases}$$

and can now state

THEOREM 21. Let $F(x)$ be an irreducible third order P-polynomial, and let

$$R_{1,1}, R_{2,2}, \ldots, R_{n-2,n-2}$$

be any $n-2$ of the $n-1$ polynomials

$$P_{0,0}, \ldots, P_{k-1,k-1}, P_{k+1,k+1}, \ldots, P_{n-1,n-1}; \qquad (k=1,\ldots,n-2)$$

then the JPA of $a^{(0)}$ with the components

$$a_i^{(0)} = R_{1,n-1-i}P_{k,k} \ (i=1,\ldots,n-2); \quad a_{n-1}^{(0)} = R_{1,1}$$

with the T-function

(6.46) $$b_s^{(v)} = a_s^{(v)}(D_k), \qquad (s=1,\ldots,n-1;\ v=0,1,\ldots)$$

is purely periodic; the length of the primitive period is $m = n(n-1)$ (n B-fugues of the form (6.39)) for $d > 1$ and $m = n-1$ (one B-fugue of the form (6.39)).

It is now obvious that Theorem 20 is a special case of Theorem 21, with the outer T-function of Theorem 20 replaced by the T-function (6.46).

Calculating the units from the periods of the periodic JPA's of Theorem 20 and Theorem 21, we can now state:

THEOREM 22. Let $F(x)$ be a third order P-polynomial with the property (6.44); let w be its greatest root; then the numbers

$$\varepsilon_k = \frac{(w - D_k)^n}{d} \qquad (k=0,\ldots,n-1) \tag{6.47}$$

are n different units of the algebraic number field $Q(w)$.

It is easily seen that one of the n units, say ε_{n-1}, is expressible as a product of the remaining n-1 units; from

$$F(w) = (w - D_0)(w - D_1) \ldots (w - D_{n-1}) - d = 0$$

we obtain

$$\frac{d}{w - D_{n-1}} = (w - D_0)(w - D_1) \ldots (w - D_{n-2}),$$

$$\frac{d^n}{(w - D_{n-1})^n} = (w - D_0)^n (w - D_1)^n \cdots (w - D_{n-2})^n,$$

$$\frac{d}{(w - D_{n-1})^n} = \frac{(w - D_0)^n}{d} \cdot \frac{(w - D_1)^n}{d} \cdots \frac{(w - D_{n-2})^n}{d}$$

$$\varepsilon_{n-1} = (\varepsilon_0 \varepsilon_1 \cdots \varepsilon_{n-2})^{-1}. \tag{6.48}$$

Since $F(x)$ is irreducible by Theorem 22 and has n real roots, the field $Q(w)$ has n-1 fundamental units; here it will only be shown that the n-1 units ε_k $(k=0,1,\ldots,n-2)$ are independent; whether they are fundamental or not, has not been proved for the general case. We prove

THEOREM 23. Let the conditions of Theorem 22 be fulfilled; then any n-1 of the n units (6.47) are independent.

Proof. Let the n roots of F(x) be arranged in decreasing magnitude

(6.49) $$w = w^{(0)} > w^{(1)} > w^{(2)} > \cdots > w^{(n)}.$$

The relative position of these roots between the outside of the sequence

$$D_0 > D_1 > \cdots > D_{n-1}$$

is such that, in virtue of $D_v \equiv D_0 \pmod d$, the inequalities hold, for every fixed v,

(6.50) $$\left|w^{(v)} - D_m\right| > \begin{cases} d & \text{for all } m \neq v \text{ except possibly one} \\ \frac{1}{2}d & \text{for the possible exception } m \neq v. \end{cases}$$

The possible exception occurs for one of the two D_m which include $w^{(v)}$ (so far $v > 0$ and for even n also $v < n-1$), and hence only for $n \geqslant 3$ (since for $n = 2$ both roots $w^{(0)}$, $w^{(1)}$ are excluded by D_0, D_1). It is obvious that in (6.47) any of the $w^{(v)}$ may take the place of w to obtain the units (in the field $Q(w^{(v)})$)

(6.51) $$e_m^{(v)} = \frac{(w^{(v)} - D_m)^n}{d}.$$

From (6.50), (6.51) we obtain, for every fixed v,

(6.52) $$\left|e_m^{(v)}\right| > \begin{cases} d^{n-1}/d = d^{n-2} & \text{for all } m \neq v \text{ except possibly one} \\ \frac{1}{2}d^{n-1}/d = \frac{1}{2}d^{n-2} & \text{for the possible exception } m \neq v. \end{cases}$$

Since the exception does not occur for $n = 2$, and since in virtue of the preposition $D_{2k-1} - D_{2k} \geqslant 2$ which we shall make in case $d = 1$, we obtain from (6.52)

(6.53) $$\left|e_m^{(v)}\right| > 1, \quad \text{for } m \neq v.$$

But

$$\prod_{m=0}^{n-1} e_m = \prod_{m=0}^{n-1} d^{-1}(w - D_m)^n$$

$$= \frac{((w - D_0)(w - D_1)\cdots(w - D_m))^n}{d^n} = \frac{d^n}{d^n} = 1,$$

so that from (6.53) it follows that

$$\left| e_v^{(v)} \right| < 1.$$

Since F(w) is irreducible, for each fixed m the $e_m^{(v)}$ are the algebraic conjugates of e_m. Therefore, by a well-known theorem of Minkowski (see H. Hasse, Zahlentheorie, Berlin 1963; 28, 2, Hilfsatz) this implies that for any fixed pair m_0, v_0 the determinant

$$\left| \log \left| e_m^{(v)} \right| \right|_{m \neq 0; v \neq 0} \neq 0,$$

so that every n-1 of the n units e_m are independent.

Example 10. Let the third order P-polynomial be

$F(x) = (x - 20)(x - 6)(x - 2)(x + 4) - 2.$

$F(w) = 0;\ 20 < w < 21;$

$D_0 = 20;\ D_1 = 6;\ D_2 = 2;\ D_3 = -4;$

here (6.44) is fulfilled; F(x) is irreducible.

Since $w^4 = 24w^3 - 60w^2 - 448w + 962$, we calculate easily

$$\frac{(w - D_1)^4}{d} = \frac{(w - 6)^4}{2} = 78w^2 - 656w + 1120;$$

$$\frac{(w - D_2)^4}{d} = \frac{(w - 2)^4}{2} = 8w^3 - 18w^2 - 240w + 489,$$

$$\frac{(w - D_3)^4}{d} = \frac{(w + 4)^4}{2} = 20w^3 + 18w^2 - 96w + 609.$$

This is the proper place to remark that most of the units found so far are in the ring of Q(w); but there are also exceptions, viz. if $w = \sqrt[n]{D^n \pm d^r D}$, then

$$\varepsilon = \frac{(w - D)^n}{d^r D}, \qquad r > 0$$

are not in the ring.

§5. Units NOT from the Period of JPA

So far we have calculated units of an algebraic number field from the period of a periodic JPA. The field in question was mainly of the form $Q(\alpha)$, $\alpha = \sqrt[n]{D^n \pm d^r D}$, and a few other cases, actually related to $Q(\alpha)$. When α is not of the above structure, the question whether or not the JPA of a properly chosen $a^{(0)}(\alpha)$, associated with any T-function, becomes periodic can only be decided upon by calculations, whereby it is mostly convenient to choose the outer T-function. This author has initiated a program with the IBM 360-Computer (precision 16 decimal places) at the Illinois Institute of Technology and tested the JPA with the outer T-function of vectors (w, w^2), $w = \sqrt[3]{m}$, for $m = 2,\ldots,1000$; also the vectors (w,w^2,w^3), $w = \sqrt[4]{m}$, and (w, w^2, w^3, w^4), $w = \sqrt[5]{m}$ were tested for $m = 2,3,\ldots,$ 100. At the end of this chapter a table is included; in many cases the JPA became periodic, with usually very long pre-periods and periods. **The w -s are irrationals of degrees 3 , 4 and 5 respectively.**

When periodicity did not occur the blame is rather to be put on the limited precision of the machine; for the author conjectures that **in many more caes ,though not in all,the JPA with the outer T-function may** becomes periodic. But these calculations also revealed that it is possible to obtain a unit even if the JPA is not periodic, and if it is - from the pre-period thus saving the time-consuming operation of trying to calculate the period.

We return to formulas (1.15), viz.

$$D_v = \begin{vmatrix} 1 & A_0^{(v+1)} & \cdots & A_0^{(v+n-1)} \\ a_1^{(0)} & A_1^{(v+1)} & \cdots & A_1^{(v+n-1)} \\ \vdots & \vdots & & \vdots \\ a_{n-1}^{(0)} & A_{n-1}^{(v+1)} & \cdots & A_{n-1}^{(v+n-1)} \end{vmatrix} = \left(\sum_{j=0}^{n-1} a_j^{(v)} A_0^{(v+j)} \right)^{-1},$$

and (1.16)

$$\prod_{i=1}^{v} a^{(i)} = \sum_{j=0}^{n-1} a_j^{(v)} A_0^{(v+j)}. \qquad (v=1,2,\ldots)$$

From (1.15) we learn the following facts: if the determinant D_v is an algebraic integer, so is $\left(\sum_{j=0}^{n-1} a_j^{(v)} A_0^{(v+j)} \right)^{-1}$; and if $\sum_{j=0}^{n-1} a_j^{(v)} A_0^{(v+j)}$ is an algebraic integer, then it is a unit. These two conditions, namely that D_v and $\sum_{j=0}^{n-1} a_j^{(v)} A_0^{(v+j)}$ be both algebraic integers, could then serve as a criterion for such units, for any v. But the tests of integrality for these two numbers are much too complicated to serve any practical purpose. Instead, the following reasoning will serve the purpose much better: suppose the JPA in question is associated with such a T-function that the components of all $b^{(v)}$ $(v=0,1,\ldots)$ be all rational integers. Then all the $A_i^{(v)}$ $(i=0,\ldots,n-1;\ v=0,1,\ldots)$ are. Suppose further that, for a fixed v, the components of $a^{(v)}$ $(v=1,2,\ldots)$ are algebraic integers; if also the components of $a^{(0)}$ are algebraic integers, then both D_v and D_v^{-1} are algebraic integers, and we have thus obtained

THEOREM 24. Let $a^{(0)} \in E_{n-1}$ be a vector whose components are algebraic integers; let the JPA of $a^{(0)}(w)$ be associated with a T-function such that the components of $b^{(v)}$ $(v=0,1,2,\ldots)$ be rational integers; if for a fixed $v \geqslant 1$ the components of $a^{(v)}$ are algebraic integers, then

(6.54) $$\varepsilon_v = \sum_{j=0}^{n-1} a_j^{(v)} A_0^{(v+j)} = \prod_{i=1}^{v} a_{n-1}^{(i)}$$

is a unit in the field Q(w).

If v from (6.54) is such that $v < \ell$ (ℓ is length of pre-period), then the unit ε_v is already calculated from the pre-period. Of course, (6.54) can hold for a set of values of v. If $v' = s + v$ $(s > 0)$ gives also a unit, then $\varepsilon_{s+v}: \varepsilon_v = \prod_{i=v+1}^{s} a_{n-1}^{(i)}$. If the **field in question is cubic and not totally real , so that there is only one fundamental unit ,then $\varepsilon_{v'}$ and ε_v are ,of course, dependent. The author was unable to decide this question.**

The table below contains some of the results obtained from the computer as mentioned before. It gives the units in a cubic field $Q(\sqrt[3]{m})$, $1 < m < 1000$. The first entries are values of m; the second entries are the lengths of the pre-period of the JPA with the outer T-function of (w, w^2), where $w = \sqrt[3]{m}$; the third entries are the lengths of the period, here denoted by L; the fourth entries contains the values of v from (6.54) and the values of ε_v^{-1}; these are taken instead of the values of ε_v, since the coefficients of ε_v^{-1} are much smaller; in this entry (a,b,c) stands for $aw^2 + bw + c$; the fifth entries contain the inverse of the unit calculated from the period and denoted by ε_p^{-1}; the sixth entries show the functional connection between ε_p and ε_v.

Table of Units in Cubic Fields Q($\sqrt[3]{m}$), 1 < m < 100

1	2	3		4	5	6
m	ℓ	L	v	ε_v^{-1}	ε_p^{-1}	$\varepsilon_p = f(\varepsilon_v)$
3	2	2	2	(1,0,-2)	(1,0,-2)	$\varepsilon_2 = \varepsilon_p$
4			13	(50,-80,1)		
11			3	(-2,4,1)		
12	2	11	12	(6,33,-107)	(6,33,-107) $=(-\frac{3}{2},3,1)^2$	$\varepsilon_p = \varepsilon_{12}$
13			5	(2,-3,-4)		
15			10	(12,-30,1)		
16	30	18	10	(-20,50,1)		
17	32	61	24	(95256,-1222479, -314856)		
18	9	9	6	(1,-3,1)	(11,12,-107)	$\varepsilon_p = \varepsilon_p^2$
19			8	(0,3,-8)		
20			8	(0,7,-19)		
21			8	(4,6,-47)		
22			3	(-4,3,23)		
24			9	(3,-9,1)		
31	5	13			(-20,54,-367)	
35			12	(16,-45,-24)		
37			3	(0,-3,10)		
38			12	(-3,55,-151)		
39			5	(2,0,-23)		
42			13	(12,-42,1)		
43	5	4	4	(0,2,-7)	(0,2,-7)	$\varepsilon_p = \varepsilon_v$
45			16	(-104,66,1081)		
46			14	(309,48,-4139)		
52	10	10	6	(1,-4,1)	(18,44,-415)	$\varepsilon_p = \varepsilon_6^2$
54			12	(20,-33,-161)		
55	13	20				
56	7	10			(39,114,-1007)	
57			10	(-88,57,1084)		
58	7	6	6	(2,-8,1)	(2,-8,1)	$\varepsilon_p = \varepsilon_6$
62	4	8	9	(6,-24,1)	(6,-24,1)	$\varepsilon_p = \varepsilon_9$
70			3	(-2,8,1)		
72	2	11	11	(33,174,-1295)	(33,174,-1295)	$\varepsilon_p = \varepsilon_{11}$
74			14	(60,-23,-961)		
86			4	(-1,6,-7)		
			11	(50,2,-983)	$\varepsilon_{11} = \varepsilon_4^2$	
90		10		(12,-54,1)		
91	4	3	3	(0,-2,9)	(0,-2,9)	$\varepsilon_p = \varepsilon_3$
98			22	(-2939,378,60729)		
106			18	(-288,3177,-8585)		
112			16	(26,-962,4033)		
129			19	(-3076,10401,25978)		
148	12	15	11	(-49,259,1)		
152			11	(-54,288,1)		
153	5	22	4	(1,4,-50)		
162			18	(-95,696,-971)		
164			10	(-15,22,329)		
165			10	(12,-66,1)		
166	4	9	9	(44,-242,1)	(44,-242,1)	$\varepsilon_p = \varepsilon_9$
168			4	(1,3,-47)		
171			10	(1,-16,58)		
182	6	39	8	(0,3,-17)		
			17	(9,-102,289)	$\varepsilon_{17} = \varepsilon_8^2$	
198			6	(1,-6,1)		

1	2	3		4	5	6
m	ℓ	L	v	ε_v^{-1}	ε_p^{-1}	$\varepsilon_p = f(\varepsilon_v)$
199			20	(-104,3216,-15231)		
204	7	13				
207	7	6	6	(2,-12,1)	(2,-12,1)	$\varepsilon_p = \varepsilon_6$
209			7	(6,81,-692)		
212			18	(738,4239,-51515)		
225			3	(-2,12,1)		
228			15	(78,531,-6155)		
230			23	(4395,-3954,-140759)		
231	6	7	7	(-2,39,-164)	(-2,39,-164)	$\varepsilon_p = \varepsilon_7$
236			9	(8,-549,3095)		
248	12	9	11	(5,27,-367)		
254	4	3	3	(0,-3,19)	(0,-3,19)	$\varepsilon_p = \varepsilon_3$
270			12	(191,-984,-1619)		
273			10	(12,-78,1)		
275	5	12	12	(-52,338,1) $= \frac{1}{3}(0,2,-13)^2$		
209	9	27			$(\frac{9}{4},-30,100)$	
305			9	(16,-15,-624)		
322			6	(1,-7,1)		
340	4	9	9	(7,-49,1)	(7,-49,1)	$\varepsilon_g = \varepsilon_p$
360			9	(-104,66,1081)		
370			15	(-144,-3141,29971)		
420			10	(12,-90,1)		
422	5	4	4	(0,2,-15)	(0,2,-15)	$\varepsilon_p = \varepsilon_4$
423			14	(-20,150,1)		
427			15	(-535,892,23620)		
464	31	9				
468			10	(-1,20,-95)		
500	7	6	6	(2,-16,1)	(2,-16,1)	$\varepsilon_p = \varepsilon_6$
506	10	18	9	(4,-32,1)		
508	4	9	9	(6,-48,1)		
509	4	9	9	(8,-64,1)		
561			10	(48,-396,1)		
612			10	(12,-102,1)		
614	4	3	3	(0,-2,17)	(0,-2,17)	$\varepsilon_4 = \varepsilon_p$
618			6	(1,-3,-47)		
635			13	(0,-42,361)		
650			12	(27,-234,1)		
651			8	(0,3,-26)		
657			12	(1,-34,220)		
665			14	(-86,-807,13596)		
703			11	(10,15,-924)		
721			12	(30,36,-2735)		
741			8	(-4,174,-1247)		
755			13	(-8,3,636)		
763			16	(-8,-733,7366)		
770			16	(108,-990,1)		
794	11	18	6	(3,-24,-35)		
812	13	22	10	(27,-252,1)		
813	4	3	3	(0,-3,28)	(0,-3,28)	$\varepsilon_p = \varepsilon_3$
855			10	(12,-114,1)		
857	4	9	9	(76,-722,1)		
914			6	(-1,-4,133)		
965	9	20				
970			6	(1,-10,1)		
985	7	6	6	(2,-20,1)		
994	10	17	9	(5,-50,1)		

For fields of higher degree the following units were calculated:

1) $w = \sqrt[4]{3}$; $a^{(0)} = (w^2, w, w^3)$;

$e_4^{-1} = 2 - w^2$; $e_6^{-1} = 1 + w - w^2$;

2) $w = \sqrt[4]{5}$; $a^{(0)} = (w, w^2, w^3)$;

$e_2 = w^2 + 2$; $e_4^{-1} = 3 - 2w$;

3) $w = \sqrt[4]{7}$; $a^{(0)} = (w, w^3, w^2)$;

$e_3^{-1} = w^2 - w - 1$; $e_5^{-1} = 8 + 2w - w^2 - 2w^3$;

4) $w = \sqrt[4]{12}$; $a^{(0)} = (w, w^2, w^3)$;

$e_8 = 15w^3 + 28w^2 + 52w + 97$;

$w = \sqrt[4]{12}$; $a^{(0)} = (w^2, w^3, w)$;

$e_4 = 2w^2 + 7$;

5) $w = \sqrt[5]{5}$; $a^{(0)} = (w, w^2, w^3, w^4)$;

$e^{-1} = 1 + w^3 - w^4$;

6) $w = \sqrt[5]{7}$; $a^{(0)} = (w, w^2, w^3, w^4)$;

$e = 4643w^4 + 6852w^3 + 10112w^2 + 14923w + 22023.$

We shall conclude this chapter with a word about the fundamentality of the units

$$\varepsilon = \frac{(w - D)^n}{w^n - D^n} \text{ in the field } Q(w).$$

$$w = \sqrt[n]{D^n \pm k} .$$

The author conjectures that these are always fundamental units, with at most a finite number of exceptions. These exceptions most likely will occur, when the ring of integers of Q(w) has an integral basis different from 1, w, $w^2, \ldots, w^{n-1}$. Very little is known about such integral bases beyond $n = 3$. Already Dedekind has shown that for the field $Q(\sqrt[3]{m})$, m a natural number, not containing cubic factors

$$m = ab^2,\ (a,b) = 1;\quad \bar{m} = a^2 b,$$

$$w = \sqrt[3]{m},\quad \bar{w} = \sqrt[3]{\bar{m}},$$

$1, w, \bar{w}$ is an integral basis, whereby $m \not\equiv 1 \pmod 9$, while $\frac{1}{3}(1 + aw + b\bar{w}),\ w, \bar{w}$ is an integral basis for $m \equiv 1 \pmod 9$. Beyond $n = 3$, the efforts to find an integral basis seem unsurmountable. A significant progress in the problem of the fundamental unit of a **cubic ,not totally real field,with one fundamental unit,has recently been** achieved by H. J. Stender in Köln [26] who proved, indeed, the conjecture of the author for the case $n = 3$, for the cases

(i) $w = \sqrt[3]{D^3 + d},\quad d \mid D;$

(ii) $w = \sqrt[3]{D^3 + 3d},\ d \mid D,\quad 3d \leqslant D;$

(iii) $w = \sqrt[3]{D^3 + 3D},\quad D \geqslant 2;$

(iv) $w = \sqrt[3]{D^3 - d},\quad d \mid D,\quad 4d \leqslant D;$

(v) $w = \sqrt[3]{D^3 - 3d},\ d \mid D,\ 12d \leqslant D.$

The reader will note that the restriction imposed on D in (ii) - (v) are a result of the periodicity of the JPA with the outer T-function of $a^{(0)} = (w, w^2)$. In the cases (i) and (iv) the unit ε takes the form, for $d = 1$,

$$\varepsilon = (w - D)^3,$$

and since $w - D$ is already in $Q(w)$, a unit of $Q(w)$ is given by $\bar{\varepsilon} = w - D$. On this case previous results by T. Nagell are known. He proved that in the cubic field $Q(w)$; $w = \sqrt[3]{D^3 \pm 1} > 0$; w^3 contains no cubic factors, the unit $\bar{\varepsilon} = w - D$ is fundamental with the only exception of $w = \sqrt[3]{28}$, where

$$\varepsilon' = \frac{1}{3}(-1 - w + \frac{1}{2}w^2),\quad (\varepsilon'^2 = w - 3 = \bar{\varepsilon})$$

is a fundamental unit. Stender's results can be summarized in

THEOREM 24. In the field $Q(w)$, w from (i) to (v) the unit

$$\varepsilon = \frac{(w - D)^3}{w^3 - D^3}, \qquad \left| w^3 - D^3 \right| > 1$$

is always fundamental, with the only exception in (i) for $D = d = 2$, $w = \sqrt[3]{10}$, where $\frac{1}{3}(-7 - w + 2w^2) = \sqrt{\varepsilon}$ is the fundamental unit.

Chapter 7

DIOPHANTINE EQUATIONS

In this chapter we shall solve Diophantine equations of any degree $n \geqslant 2$; they are all homogeneous in n variables and represent a generalization of the Pell equation to higher dimensions. In certain cases infinitely many solution vectors will be stated explicitly. All the solutions of the Diophantine equations investigated here are derived from periodic JPA's of algebraic vectors in E_{n-1}.

§ 1. Explicit Summation Formulas

Of paramount importance for the investigation of the properties of a periodic JPA is the calculation of the $A_i^{(v)}$ $(i=0,\ldots,n-1;$ $v=0,1,\ldots)$ by means of the components of the vectors $b^{(v)}$. The solution of this problem in its widest generality, seems to be very difficult. It has been solved by H. Hasse and the author in [3,c] for a special case, and most of the periodic JPA's dealt with in this monograph fall into that category.

THEOREM 25. Let a JPA be purely periodic with length of period $= n$, and let the vectors of the period have the form (3.11), viz.

$$b^{(0)} = (k_1,\ldots,k_{n-1}),$$

$$b^{(i)} = (k_1,\ldots,k_{n-1-i},k'_{n-i},\ldots,k'_{n-1}), \quad (i=1,\ldots,n-2)$$

$$b^{(n-1)} = (k'_1,\ldots,k'_{n-1}),$$

$$k'_j = d^{-1}k_j \quad (j=1,\ldots,n-1);\ d,k_j \in Q;\ d \neq 0.$$

Then the numbers $A_0^{(v)}$ $(v=n,n+1,\ldots)$ calculated from the components of the $b^{(v)}$ $(v=0,\ldots,n-1)$ by means of (1.8) and $b^{(sn+j)} = b^{(j)}$ $(s=1,2,\ldots;\ j=0,\ldots,n-1)$ are given by the formula

$$(7.1)\quad \begin{cases} A_0^{((s+1)n+j)} = \\ d^s \sum\limits_{L=sn+j} \begin{pmatrix} x_0+x_1+\cdots+x_{n-1} \\ x_0,\ldots,x_{n-1} \end{pmatrix} d^{-1(x_0+\cdots+x_{n-1})}\ k_1^{\ x_1}\ldots k_{n-1}^{\ x_{n-1}}, \\ L = xn_0+(n-1)x_1+\cdots+2x_{n-2}+1\cdot x_{n-1}, \\ \qquad x_i\ (i=0,\ldots,n-1) \text{ non-negative integers}, \\ \begin{pmatrix} x_0+x_1+\cdots+x_{n-1} \\ x_0,\ldots,x_{n-1} \end{pmatrix} \equiv \dfrac{(x_0+x_1+\cdots+x_{n-1})!}{x_0!\ x_1!\ \cdots\ x_{n-1}!}\ . \end{cases}$$

The proof of Theorem 25 is quite cumbersome and will be omitted here. The tools used in proving Theorem 25 are borrowed from Euler's generating functions, and we shall make use of them, at a later stage, for some special cases.

As can be seen from (7.1), one must first solve the Diophantine equation

$$nx_0+(n-1)x_1+\cdots+2x_{n-2}+1\cdot x_{n-1} = sn+j,$$

and we shall now demonstrate the technique of calculating the $A_0^{(v)}$ for a special case.

Example 11. Let $n = 3$; we calculate successively by means of (1.8), the period having the form

$$b^{(0)} = (k_1,\ k_2),$$
$$b^{(1)} = (k_1,\ d^{-1}k_2),$$
$$b^{(2)} = (d^{-1}k_1,\ d^{-1}k_2),$$

and with the notation $d^{-1} = t$,

$$A_0^{(3)} = 1;\quad A_0^{(4)} = tk_2;\quad A_0^{(5)} = t^2k_2^{\ 2}+tk_1;$$
$$A_0^{(6)} = t^2k_2^{\ 3}+ 2tk_2k_1 + 1;$$
$$A_0^{(7)} = t^3k_2^{\ 4}+ 3t^2k_2^2k_1 + 2tk_2 + tk_1^{\ 2};$$
$$A_0^{(8)} = t^4k_2^{\ 5}+ 4t^3k_2^{\ 3}k_1 + 3t^2k_2^{\ 2} + 3t^2k_2k_1^{\ 2}+ 2tk_1.$$

We shall test $A_0^{(6)}$, $A_0^{(7)}$, $A_0^{(8)}$; we obtain from (7.1) for $s = 1$, $j = 0$,

$$A_0^{(6)} = d \sum_{L=3} \binom{x_0+x_1+x_2}{x_0,x_1,x_2} d^{-(x_0+x_1+x_2)} k_1^{x_1} k_2^{x_2}$$

$3x_0 + 2x_1 + x_2 = 3$ has the solutions

(1,0,0); (0,1,1); (0,0,3);

therefore

$$A_0^{(6)} = d \left[\binom{1}{1} d^{-1} k_1{}^0 k_2{}^0 + \binom{2}{1} d^{-2} k_1 k_2 + \binom{3}{3} d^{-3} k_2{}^3 \right]$$

$$= 1 + 2tk_1k_2 + t^2 k_2{}^3;$$

for $s = 1$, $j = 1$,

$$A_0^{(7)} = d \sum_{L=4} \binom{x_0+x_1+x_2}{x_0,x_1,x_2} d^{-(x_0+x_1+x_2)} k_1^{x_1} k_2^{x_2};$$

$3x_0 + 2x_1 + x_2 = 4$ has the solutions

(1,0,1); (0,2,0); (0,1,2); (0,0,4);

therefore

$$A_0^{(7)} = d \left[\binom{2}{1} d^{-2} k_2 + \binom{2}{2} d^{-2} k_1{}^2 + \binom{3}{2} d^{-3} k_1 k_2{}^2 + \binom{4}{4} d^{-4} k_2{}^4 \right]$$

$$= 2tk_2 + tk_1{}^2 + 3t^2 k_2{}^2 k_1 + t^3 k_2{}^4;$$

for $s = 1$, $j = 2$,

$$A_0^{(8)} = d \sum_{L=5} \binom{x_0+x_1+x_2}{x_0,x_1,x_2} d^{-(x_0+x_1+x_2)} k_1^{x_1} k_2^{x_2};$$

$3x_0 + 2x_1 + x_2 = 5$ has the solutions

(1,1,0); (1,0,2); (0,2,1); (0,1,3); (0,0,5);

Therefore

$$A_0^{(8)} = d \left[\binom{2}{1} d^{-2} k_1 + \binom{3}{2} d^{-3} k_2^2 + \binom{3}{2} d^{-3} k_1^2 k_2 + \binom{4}{3} d^{-4} k_1 k_2^3 + \binom{5}{5} d^{-5} k_2^5 \right]$$

$$= 2tk_1 + 3t^2 k_2^2 + 3t^2 k_1^2 k_2 + 4t^3 k_1 k_2{}^3 + t^4 k_2^5.$$

Now let F(x) be a second order P-polynomial and w its root as defined by (5.5). By (5.7) w is calculated from

$$w = \lim_{s\to\infty} \frac{A_0^{((s+1)n)}}{A_0^{((s+1)n+1)}},$$

and using (7.2), we obtain

$$(7.2)\qquad w = \lim_{s\to\infty} \frac{\sum\limits_{L=sn} \binom{x_0+x_1+\cdots+x_{n-1}}{x_0,x_1,\ldots,x_{n-1}} d^{-(x_0+\cdots+x_{n-1})} k_1^{x_1} k_2^{x_2} \cdots k_{n-1}^{x_{n-1}}}{\sum\limits_{L=sn+1} \binom{x_0+x_1+\cdots+x_{n-1}}{x_0,x_1,\ldots,x_{n-1}} d^{-(x_0+\cdots+x_{n-1})} k_1^{x_1} k_2^{x_2} \cdots k_{n-1}^{x_{n-1}}}.$$

We shall now investigate a few specifications of the components of the vector $b^{(0)} = (k_1, k_2, \ldots, k_{n-1})$ and start with

$$(7.3)\qquad k_1 = k_2 = \cdots = k_{n-2} = 0;\quad d^{-1}k_{n-1} = b.$$

The formulas for calculating the $A_0^{(v)}$ become

$$(7.4)\qquad \begin{aligned} A_0^{(v+n)} &= A_0^{(v)} + d^{-1}k_{n-1}A_0^{(v+n-1)}, && v \not\equiv 0 \pmod n; \\ A_0^{(v+n)} &= A_0^{(v)} + k_{n-1}A_0^{(v+n-1)}, && v \equiv 0 \pmod n. \end{aligned}$$

We now use the following generating function for the calculation of $A_0^{(v+n)}$.

$$(7.5)\qquad \frac{1}{d - k_{n-1}x - x^n} = \sum_{v=0}^{\infty} d^{-\left[\frac{v+n}{n}\right]} A_0^{(v+n)} x^v,$$

and leave it to the reader to verify that comparison of coefficients in

$$1 = \sum_{v=0}^{\infty} d^{-\left[\frac{v+n}{n}\right]} A_0^{(v+n)} (dx^v - k_{n-1}x^{v+1} - x^{n+v})$$

indeed yields the recurrency relation for the $A_0^{(v+n)}$ as demanded in (7.4). We now obtain

$$\frac{1}{d - k_{n-1}x - x^n} = \frac{d^{-1}}{1 - bx - d^{-1}x^n}$$

$$= d^{-1}\left[1 - (bx + d^{-1}x^n)\right]^{-1} = d^{-1}\sum_{i=0}^{\infty} (bx + d^{-1}x^n)^i .$$

In the expression $\sum_{i=0}^{\infty} (bx + d^{-1}x^n)^i$ we open brackets and collect equal powers of x; the reader will again verify without difficulty that the coefficient of

$$\text{(7.6)} \quad \begin{cases} x^{sn+f}, \quad (s=0,1,\ldots; \ f=0,1,\ldots,n-1) \\ \text{equals} \\ \sum_{i=0}^{s} \binom{sn+f-i(n-1)}{i} d^{-i} b^{sn+f-in}. \end{cases}$$

Comparison of coefficients in (7.5) thus yields, for $v = sn+f$,

$$d^{-\left[\frac{sn+f+n}{n}\right]} A_0^{(sn+f+n)} = d^{-1} \sum_{i=0}^{s} \binom{(s-i)n+f+i}{i} d^{-i}\, b^{(s-i)n+f},$$

and, taking into account that $\left[\frac{sn+f+n}{n}\right] = s + 1$, for $f = 0,\ldots,n-1$

$$A_0^{((s+1)n+f)} = b^f \sum_{i=0}^{s} \binom{(s-i)n+f+i}{i} d^{s-i}\, b^{(s-i)n}.$$

Denoting, with $b = d^{-1}k_{n-1}$, $b^n = z$; $dz = x$, we finally obtain

$$\text{(7.7)} \qquad A_0^{((s+1)n+f)} = b^f \sum_{i=0}^{s} \binom{(s-i)n+f+i}{i} x^i . \quad (s=0,1,\ldots;f=0,\ldots,n-1)$$

Now let F(x) be a second order P-polynomial of the form

$$F(x) = x^n + k_{n-1}x - d$$

and w its root as defined by (5.5). From (5.7), w is calculated by

$$w = \lim_{s\to\infty} \frac{A_0^{((s+1)n)}}{A_0^{((s+1)n+1)}} ,$$

so that in view of (7.7), for $f = 0$,

$$w = \frac{1}{b} \lim_{s \to +\infty} \frac{\sum_{i=0}^{s} \binom{(s-i)n+i}{i} x^i}{\sum_{i=0}^{s} \binom{(s-i)n+i+1}{i} x^i}, \tag{7.8}$$

$$b = d^{-1}k_{n-1}; \; x = db^n.$$

The reader will note that (7.8) solves the cubic equation completely, if this can be brought to the form $x^3 + px + q = 0$ with $0 < |x| < 1$.

Example 12. We shall solve

$$F(z) = z^3 - 9z^2 + 15z + 11 = 0;$$

we obtain: $F(3) > 0$, $F(4) < 0$; by means of the substitution $z = x + 3$ we obtain

$$F(x) = x^3 - 12x + 2; \quad F(w) = 0; \quad 0 < w < 1,$$

$F(x)$ is a second order P-polynomial and irreducible; from (7.8) we obtain

$$b = 6; \quad x = -2\cdot 6^3 = -432;$$

$$w = \frac{1}{6} \lim_{s \to \infty} \frac{\sum_{i=0}^{s} (-1)^i \binom{3s-2i}{i} 432^i}{\sum_{i=0}^{s} (-1)^i \binom{3s-2i+1}{i} 432^i}.$$

A very interesting case is the following specification of the k_i and d

$$d = k_1 = \cdots = k_{n-1} = 1. \tag{7.9}$$

The formula for calculating the $A_0^{(v)}$ becomes

$$A_0^{(v+n)} = A_0^{(v)} + A_0^{(v+1)} + \cdots + A_0^{(v+n-1)}. \tag{7.10}$$

In view of (7.9), formula (7.1) takes the form

$$(7.11)\qquad A_0^{((s+1)n+k)} = \sum_{nx_0+(n-1)x_1+\cdots+x_{n-1}=sn+k} \frac{(x_0+x_1+\cdots+x_{n-1})!}{x_0!\,x_1!\cdots x_{n-1}!}\,.$$

In a recent paper [2,i] the following formula was proved by induction

$$(7.12)\qquad A_0^{((s+1)n+k)} = 2^{k-s-1}\sum_{j=0}^{s}(-1)^j\left[\binom{(s-j)n+k}{j}+\binom{(s-j)n+k-1}{j-1}\right]2^{(s-j)(n+1)}$$

$$(k=1,\ldots,n).$$

Comparing the two expressions for $A_0^{((s+1)n+k)}$ we obtain the interesting identity

$$(7.13)\quad \begin{cases} \displaystyle\sum_{L=sn+k}\binom{x_0+x_1+\cdots+x_{n-1}}{x_0,x_1,\ldots,x_{n-1}} = \\ \displaystyle 2^{k-s-1}\sum_{j=0}^{s}(-1)^j\left[\binom{(s-j)n+k}{j}+\binom{(s-j)n+k-1}{n-1}\right]2^{(s-j)(n+1)}. \\ \qquad\qquad (s=0,1,\ldots;\ k=1,\ldots,n) \\ \displaystyle L=\sum_{i=0}^{n-1}(n-i)x_i;\quad x_i \text{ non-negative integers;} \\ \displaystyle \binom{x_0+x_1+\cdots x_{n-1}}{x_0,\ldots,x_{n-1}} = \frac{(x_0+x_1+\cdots+x_{n-1})!}{x_0!\,x_1!\cdots x_{n-1}!};\quad \binom{v}{-1}=0. \end{cases}$$

For practical purposes, the expression for $A_0^{((s+1)n+k)}$ given by (7.12) is much more advantageous than that of (7.11). In [2,j] the author has solved a probability problem which, with formula (7.12), would now sound as follows:

The probability that a fragile stick of $sn+k$ $(s=0,1,\ldots;$ $k=1,\ldots,n)$ units length, breakable into pieces, each consisting of an integral number of units not exceeding n, would break into y_1 pieces of one unit length each, into y_2 pieces of two units length each,..., into y_n pieces of n units length each, so that

$$y_1+2y_n+\cdots+ny_n = sn+k;\quad Y_i = (y_1,\ldots,y_n)$$

is given by the formula

$$p(Y_i) =$$

$$(7.14)\qquad \frac{\binom{y_1+y_2+\cdots+y_n}{y_1,y_2,\ldots,y_n}}{2^{k-s-1}\sum_{j=0}^{s}(-1)^j\left[\binom{(s-j)n+k}{j}+\binom{(s-j)n+k-1}{j-1}\right]2^{(s-j)(n+1)}}\,.$$

In virtue of (7.2), we can now also state:

The (only) positive real root of the equation

$$x^n + x^{n-1} + \cdots + x - 1 = 0 \qquad (n \geqslant 2)$$

is given by the formula

$$(7.15)\qquad w = \lim_{s\to\infty} \frac{\sum_{j=0}^{s-1}(-1)^j\left[\binom{(s-j)n}{j}+\binom{(s-j)n-1}{j-1}\right]2^{sn-j(n+1)-1}}{\sum_{j=0}^{s}(-1)^j\left[\binom{(s-j)n+1}{j}+\binom{(s-j)n}{n-1}\right]2^{sn-j(n+1)}}\,.$$

Since $A_0^{(1)} = A_0^{(2)} = \cdots + A_0^{(n-1)} = 1$, $A_0^{(n)} = 1$, and, in this case, $A_0^{(v+n)} = A_0^{(v)} + A_0^{(v+1)} + \cdots + A_0^{(v+n-1)}$, $(v=0,1,\ldots)$ and, since the generalized Fibonacci numbers of dimension n are also defined by $F_1^{(n)} = F_2^{(n)} = \cdots = F_{n-1}^{(n)} = 0$, $F_n^{(n)} = 1$, $F_{n+v}^{(n)} = F_v^{(n)} + F_{v+1}^{(n)} + \cdots + F_{v+n-1}^{(n)}$ $(v=1,2,\ldots)$ we obtain $F_{n+v}^{(n)} = A_0^{(n+v)}$, so that formula (7.12) also serves the solution of finding the general element of the generalized Fibonacci sequence $\langle F_v^{(n)}\rangle$. (See also the author's paper [2,k].)

In [2,ℓ] the author has given a direct proof of (7.12) by means of generating functions. By the same method he also proved the formula for generalized Fibonacci numbers of dimension n = 3

$$(7.16)\qquad F^{(3)}_{3(s+1)+k} = \sum_{t=0}^{\left[\frac{3s+k}{2}\right]} \sum_{j=0}^{t} \binom{m-t-j}{t}\binom{t}{j},$$

which again yields the identity

$$(7.17)\quad \begin{cases} \displaystyle\sum_{j=0}^{\left[\frac{3s+k}{2}\right]} \sum_{j=0}^{t} \binom{m-t-j}{t}\binom{t}{j} = \\ \displaystyle 2^{k-s-1}\sum_{j=0}^{s}(-1)^j\left[\binom{3(s-j)+k}{j}+\binom{3(s-j)+k-1}{j-1}\right]2^{4(s-j)}. \\ \qquad (s=0,1,\dots;\ k=1,2,3) \end{cases}$$

In [3,c] formulas for calculating the $A_i^{(v)}$ $(i=1,\dots,n-1)$ in the case of a purely periodic JPA of length n were given similar to that for calculating $A_0^{(v)}$ in (7.1). But, as shall be demonstrated for the special case $n = 3$, the $A_i^{(v)}$ can be linearly expressed by the $A_0^{(v)}$. Let again w be a real root of a third order P-polynomial $w^3 + k_1w^2 + k_2w - d = 0$, and regard the JPA of the vector $a^{(0)} = (w + k_1,\ w^2 + k_1w + k_2)$. We then obtain

$$w + k_1 = \frac{A_1^{(3s)}+(w+k_1)A_1^{(3s+1)}+(w^2+k_1w+k_2)A_1^{(3s+2)}}{A_0^{(3s)}+(w+k_1)A_0^{(3s+1)}+(w^2+k_1w+k_2)A_0^{(3s+2)}},$$

$$\frac{d}{w} = w^2+k_1w+k_2 = \frac{A_2^{(3s)}+(w+k_1)A_2^{(3s+1)}+(w^2+k_1w+k_2)A_2^{(3s+2)}}{A_0^{(3s)}+(w+k_1)A_0^{(3s+1)}+(w^2+k_1w+k_2)A_0^{(3s+2)}}.$$

We obtain from the second equation, taking into account that $w^3 = d - k_2w - k_1w^2$,

$$d\left[A_0^{(3s)}+(w+k_1)A_0^{(3s+1)}+(w^2+k_1w+k_2)A_0^{(3s+2)}\right]$$
$$= A_2^{(3s)}w + (w^2+k_1w)A_2^{(3s+1)} + (w^3+k_1w^2+k_2w)A_2^{(3s+2)},$$

$$d\left[A_0^{(3s)}+k_1A_0^{(3s+1)}+k_2A_0^{(3s+2)}+(A_0^{(3s+1)}+k_1A_0^{(3s+2)})w+A_0^{(3s+2)}w^2\right]$$
$$= dA_2^{(3s+2)} + (A_2^{(3s)}+k_1A_2^{(3s+1)})w + A_2^{(3s+1)}w^2.$$

But w is a third degree irrational, and we obtain from the above equation, by comparison of coefficients of equal powers of w,

$$A_2^{(3s+2)} = A_0^{(3s)}+k_1A_0^{(3s+1)}+k_2A_0^{(3s+2)} = A_0^{3(s+1)},$$

$$A_2^{(3s)} + k_1A_2^{(3s+1)} = d(A_0^{(3s+1)} + k_1A_0^{(3s+2)}),$$

$$A_2^{(3s+1)} = dA_0^{(3s+2)},$$

and finally

$$(7.18)\qquad \begin{cases} A_2^{(3s)} = dA_0^{(3s+1)}; \\ A_2^{(3s+1)} = dA_0^{(3s+2)}; \\ A_2^{(3s+2)} = A_0^{(3(s+1))}. \\ (s = 0,1,\ldots) \end{cases}$$

Similarly we calculate the $A_1^{(3s)}$, $A_1^{(3s+1)}$, $A_1^{(3s+2)}$ to obtain formula (3.52). It is, of course, clear that formulas (7.18) and (3.52) hold for any values of k_1, k_2, not necessarily for the coefficients of a second order P-polynomial.

§ 2. Diophantine Equations

In this section we shall solve (partly or completely) homogeneous Diophantine equations of degree n in n variables. They can all be classified as Generalized Pell Equations of dimension $n \geqslant 2$. With one special type of these the author has dealt in [2,m]. In what sense they generalize the Pell equation $x^2 - my^2 = 1$ (m a squarefree positive rational integer) will be explained below. One could expect the generalized Pell equation to be of the form $x^n - my^n = 1$ $(n \geqslant 3)$, but that is not so. Let $w = \sqrt[n]{m}$ be an n-th degree irrational and m a natural number. We consider the field $Q(w)$; this has a basis: $1,w,\ldots,w^{n-1}$, so that any number in $Q(w)$ has the form

$$(7.19)\qquad \alpha = x_1 + x_2w + x_3w^2 + \cdots + x_nw^{n-1}. \quad (x_i \in Q;\ i = 1,\ldots,n)$$

We shall find the norm of α; to this end we find the field equation of α and its free term. We use the known technique of multiplying (7.19) successively by w, bearing in mind that $w^n = m$. We thus obtain

$$\begin{array}{l} -(\alpha-x_1) + x_2 w + x_3 w^2 + \dots + x_n w^{n-1} = 0, \\ m x_n - (\alpha-x_1) w + x_2 w^2 + \dots + x_{n-1} w^{n-1} = 0, \\ m x_{n-1} + m x_n w - (\alpha-x_1) w^2 + \dots + x_{n-2} w^{n-1} = 0, \\ \text{- -} \\ m x_2 + m x_3 w + m x_4 w^2 + \dots + m x_n w^{n-1} - (\alpha-x_1) w^{n-1} = 0. \end{array}$$

The field equation of α is then given by

$$(7.20) \qquad \begin{vmatrix} -(\alpha-x_1), & x_2, & x_3, & \dots, & x_n \\ m x_n, & -(\alpha-x_1), & x_2, & \dots, & x_{n-1} \\ m x_{n-1}, & m x_n, & -(\alpha-x_1), & \dots, & x_{n-2} \\ \dots & \dots & \dots & \dots & \dots \\ m x_2, & m x_3, & \dots, & m x_n, & -(\alpha-x_1) \end{vmatrix} = 0$$

and from (7.20) it is obvious that (since α has the coefficient $(-1)^n$)

$$(7.21) \qquad N(\alpha) = \begin{vmatrix} x_1, & x_2, & x_3, & \dots, & x_{n-1}, & x_n \\ m x_n, & x_1, & x_2, & \dots, & x_{n-2}, & x_{n-1} \\ m x_{n-1}, & m x_n, & x_1, & \dots, & x_{n-3}, & x_{n-2} \\ \dots & \dots & \dots & \dots & \dots & \dots \\ m x_2, & m x_3, & m x_4, & \dots, & m x_n, & x_1 \end{vmatrix} .$$

We shall denote

$$(7.22) \qquad N(\alpha) = D(x_1, \dots, x_n; m).$$

In the cases $n = 2,3$, this determinant, in expanded form, equals

$$(7.23) \qquad \begin{array}{l} D(x_1, x_2; m) = x_1^2 - m x_2^2; \\ D(x_1, x_2, x_3; m) = x_1^3 + m x_2^3 + m^2 x_3^3 - 3m x_1 x_2 x_3. \end{array}$$

We shall now presume that α is a unit in the ring of $Q(w)$, $\alpha \in R(w)$, so that the numbers $x_1,\ldots,x_n$ are rational integers. Since $N(\alpha) = \pm 1$, we can now state

THEOREM 26. Let $w = \sqrt[n]{m}$ be an n-th degree irrational ($n \geq 2$) and m a natural number; let

$$\alpha = x_1 + x_2 w + \cdots + x_{n-1} w^{n-2} + x_n w^n$$

be a unit in the ring $R(w)$; then the Diophantine equation

$$D(x_1,\ldots,x_n;m) = \pm 1 \tag{7.24}$$

has infinitely many solution vectors $(x_{1i}, x_{2i},\ldots,x_{ni})$, where

$$\alpha^i = x_{1i} + x_{2i} w + \cdots + x_{n-1,i} w^{n-2} + x_{n,i} w^{n-1}. \tag{7.25}$$

$$(i = \pm 1, \pm 2, \ldots)$$

The equation $D(x_1,\ldots,x_n\ ;\ m) = 1$ is always solvable; the equation $D(x_1,\ldots,x_n;\ m) = -1$ is solvable only if ,or any other unit, have norm -1.

For $n = 2$, (7.24) takes the form

$$x_1^2 - mx_2^2 = \pm 1$$

which is the Pell equation; Theorem 26 then states the properties of this equation which are all well known — with one basic exception: in the case of the Pell equation, α^i supplies all solutions of this equation, (if α is fundamental) while in the case of $D(x_1,\ldots,x_n;m) = \pm 1$ and $n > 3$ only one class of infinitely many solutions is supplied; all classes of infinitely many solutions are given by the complete set of fundamental units which are in the ring $R(w)$.

From the preceding chapter we know, that in the field

$$Q(w);\ w = \sqrt[n]{m};\ m = D^n + d;\ d|D;\ n \geq 2;\ n,D,d \text{ natural integers}$$

$$\varepsilon' = \frac{(w - D)^n}{d}$$

is a unit in $R(w)$. If we carry out the JPA with the inner T-function of the vector

$$a^{(0)} = (\ldots, \sum_{i=0}^{s} \binom{n-s-1+i}{i} D^i w^{s-i}, \ldots) \qquad (s=1,\ldots,n-1)$$

then the unit obtained from the period of this purely periodic JPA is

$$\varepsilon = \frac{d}{(w - D)^n},$$

and the formula holds

$$(7.26) \qquad \varepsilon^t = A_0^{(tn)} + a_1^{(0)} A_0^{(tn+1)} + \cdots + a_{n-1}^{(0)} A_0^{(tn+n-1)}. \quad (t=1,2,\ldots)$$

We thus obtain from (7.26)

$$\varepsilon^t = \sum_{s=0}^{n-1} a_s^{(0)} A_0^{(tn+s)}$$

$$= \sum_{s=0}^{n-1} \sum_{i=0}^{s} \binom{n-s-1+i}{i} D^i w^{s-i} A_0^{(tn+s)},$$

and hence, after easy rearrangements,

$$(7.27) \quad \begin{cases} \varepsilon^t = x_{1t} + x_{2t} w + \cdots + x_{n-1,t} w^{n-2} + x_{n,t} w^{n-1} \\ x_{s,t} = \sum_{i=0}^{n-s} \binom{n-s}{i} D^i A_0^{(tn+s-1+i)}. \\ \qquad (t=1,2,\ldots;\ s=1,2,\ldots,n) \end{cases}$$

We shall now find the norm of ε and obtain

$$(7.28) \qquad N(\varepsilon) = N\left(\frac{d}{(w - D)^n}\right) = \frac{d^n}{(N(w - D))^n}.$$

We calculate $N(w - D)$ from (7.21) and obtain, since here $x_1 = -D$, $x_2 = 1$, $x_3 = \cdots = x_n = 0$

$$N(w - D) = \begin{vmatrix} -D, & 1, & 0, & \ldots, & 0 \\ 0, & -D, & 1, & 0, \ldots, & 0 \\ - & - & - & - & - \\ 0, & 0, & \ldots, & -D, & 1 \\ m, & 0, & \ldots, & 0, & -D \end{vmatrix}$$

$$= (-D)^n + (-1)^{n-1} m = (-1)^n (D^n - m) = (-1)^n (D^n - D^n - d) = (-1)^{n-1} d.$$

$$N(w - D) = (-1)^{n-1} d \quad . \tag{7.29}$$

From (7.29) , (7.28) we thus obtain

$$N(\varepsilon) = 1 \quad . \tag{7.30}$$

For $d = 1$, $= (w - D)^{-1}$ is already a unit in $R(w)$,and we obtain,in this case, from (7.28) ,

$$N((w - D)^{-1}) = (-1)^{n-1} \quad . \qquad (d = 1) \tag{7.31}$$

For $d = 1$,we obtain from (7.30) , $N(\;_1^t) = (-1)^{(n-1)t}$, so that the equations

$$D(x_1, \ldots, x_n \; ; \; m) = -1 \quad , \quad t \text{ is odd} \; ,$$

$$D(x_1, \ldots, x_n \; ; \; m) = 1 \quad , \quad t \text{ is even} \; ,$$

are both solvable explicitly for even n 's .

If n is a composite number ,we can state more than one class of explcit solutions of $D(x_1, \ldots, x_n \; ; \; m) = 1$ .Let again be $m = D^n + d$; $d|D$; $w = \sqrt[n]{m}$; $n = sk$; then by formula (6.35) or Theorem 19, independent units in $Q(w)$ are given by the formula

$$\varepsilon^{(s)} = \frac{d}{(w^s - D^s)^k} \; , \quad 1 \leqslant s < n.$$

Similar to formula (7.26) we obtain

$$(7.32) \quad \begin{cases} \left(\varepsilon^{(s)}\right)^t = A_0'^{(tk)} + a_1^{(0)'} A_0'^{(tk+1)} + \cdots + a_{k-1}^{(0)'} A_0'^{(tk+k-1)} \\ a^{(0)'} = \left(\ldots, \sum_{i=0}^{p} \binom{k-j-1+i}{i} w'^{k-i} D^{si}, \ldots \right) \\ w' = \sqrt[k]{D^{sk} + d} = \sqrt[k]{(D^s)^k + d}. \qquad (t \geqslant 1) \\ \text{and the } A_0'^{(v)} \text{ are obtained from the JPA of } a^{(0)'}. \end{cases}$$

We further obtain

$$\left(\varepsilon^{(s)}\right)^{-t} = \frac{(w^s - D^s)^{kt}}{d^t} \; , \qquad (t \geqslant 1)$$

and explcit formulas for these units can be stated as before.

We shall now solve homogeneous Diophantine equations in n variables of degree $n \geqslant 2$ of a far more complex structure, involving n parameters, namely

$$D(x_1,\dots,x_n;\ k_1,\dots,k_{n-1};\ d) = \pm 1, \tag{7.33}$$

where $D(x_1,\dots,x_n;\ k_1,\dots,k_{n-1};\ d)$ is a determinant of the form

$$D(x_1,\dots,x_n;\ k_1,\dots,k_{n-1},\ d) =$$

$$\begin{vmatrix} x_1 & dx_n-\sum_{i=2}^{n-1}k_ix_i, & dx_{n-1}-\sum_{i=3}^{n-1}k_ix_{i-1}, & \dots, & dx_2 \\ x_2, & x_1+k_1x_2, & dx_n-\sum_{i=3}^{n-1}k_ix_i, & \dots, & dx_3 \\ x_3, & x_2+k_1x_2, & x_1+k_1x_2+k_2x_3, & \dots, & dx_4 \\ x_4, & x_3+k_1x_3, & x_2+k_1x_3+k_2x_4, & \dots, & dx_5 \\ \vdots & \vdots & \vdots & & \vdots \\ x_{n-1}, & x_{n-2}+k_1x_{n-1}, & x_{n-3}+k_1x_{n-2}+k_2x_{n-1}, & \dots, & dx_n \\ x_n, & x_{n-1}+k_1x_n, & x_{n-2}+k_1x_{n-1}+k_2x_n, & \dots, & x_1+\sum_{i=1}^{n-1}k_ix_{i+1} \end{vmatrix}. \tag{7.34}$$

Our starting point will be a second order (irreducible) P-polynomial

$$F(x) = x^n + k_1x^{n-1} + \dots + k_{n-1}x - d, \qquad (d \mid k_i)$$

with $\quad F(w) = w^n + k_1w^{n-1} + \dots + k_{n-1}w - d = 0, \quad 0 < |w| < 1$

and the purely periodic JPA with the inner T-function of the vector

$$a^{(0)} = (a_1^{(0)},\ldots,a_s^{(0)},\ldots,a_{n-1}^{(0)}),$$

$$a_s^{(0)} = \sum_{i=0}^{s} k_i w^{s-i}, \qquad (s=1,\ldots,n-1;\ k_0 = 1).$$

The reader will recall the simple formulas

$$wa_s^{(0)} = a_{s+1}^{(0)} - k_{s+1}, \quad (s=0,\ldots,n-2;\ a_0^{(0)} = 1)$$

$$wa_{n-1}^{(0)} = d\ .$$

A unit of $\Omega(w)$ has here the form

$$\varepsilon = \frac{d}{w^n} = \sum_{i=0}^{n-1} a_i^{(0)} A_0^{(n+i)}$$

$$\varepsilon^t = \left(\frac{d}{w^n}\right)^t = \sum_{i=0}^{n-1} a_i^{(0)} A_0^{(tn+i)}. \qquad (t=1,2,\ldots)$$

We could, of course, proceed, as in the previous paragraph, to find the norm of any number $\alpha \in \Omega(w)$, but the expression thus obtained, is very complicated and demands lengthy transformations to put it into a form acceptable to a mathematical eye. To solve **(7.34)** we shall therefore use a different device already hinted at, at the end **of the first paragraph of this chapter.Proceeding in the same way,we obtain from the general formula**

$$a_s^{(0)} = \frac{\sum_{i=0}^{n-1} a_i^{(0)} A_s^{(nt+i)}}{\sum_{i=0}^{n-1} a_i^{(0)} A_0^{(nt+i)}}; \qquad (s=0,\ldots,n-1)$$

(7.35)
$$\begin{cases} A_s^{(tn+q)} = dA_0^{(tn+n-s+q)} - \sum_{i=s+1}^{n-1} k_i A_0^{(tn+i-s+q)}; \\ (q=0,\ldots,s=1) \\ A_s^{(tn+s-1+j)} = \sum_{i=0}^{s} k_i A_0^{(tn+j-1+i)}. \quad (j=1,\ldots,n-s;\ k_0 = 1) \\ s = 1,\ldots,n-2. \end{cases}$$

$$(7.36)\qquad \begin{aligned} A_{n-1}^{(tn+j)} &= dA_0^{(tn+j+1)}, \qquad (j=0,\ldots,n-2)\\ A_{n-1}^{(tn+n-1)} &= \sum_{i=0}^{n-1} k_i A_0^{(tn+i)}. \end{aligned}$$

We use the determinant of the transpose of the matrices $A^{(v)}$ and obtain, for $v = tn$

$$\begin{vmatrix} A_0^{(tn)}, & A_1^{(tn)}, & \ldots, & A_{n-1}^{(tn)} \\ A_0^{(tn+1)}, & A_1^{(tn+1)}, & \ldots, & A_{n-1}^{(tn+1)} \\ \vdots & \vdots & & \vdots \\ A_0^{(tn+n-1)}, & A_1^{(tn+n-1)}, & \ldots, & A_{n-1}^{(tn+n-1)} \end{vmatrix} = (-1)^{tn(n-1)} = 1;$$

substituting in this determinant

$$(7.37)\qquad A_0^{(tn+i)} = x_{i+1}, \qquad (i=0,\ldots,n-1)$$

and for $A_i^{(tn+j)}$ $(i=1,\ldots,n-1;\ j=0,\ldots,n-1)$ the values from (7.35), (7.36) we obtain

$$(7.38)\qquad \begin{aligned} &D(x_1,\ldots,x_n;\ k_1,\ldots,k_{n-1};d) = 1\\ &x_{i+1} = A_0^{(tn+i)}. \qquad (i=0,\ldots,n-1;\ t=1,2,\ldots) \end{aligned}$$

Thus we obtain an infinity of solution vectors $X_t = (A_0^{(tn)}, A_0^{(tn+1)},\ldots,A_0^{(tn+n-1)})$ of the Diophantine equation $D(x_1,\ldots,x_n;\ k_1,\ldots,k_{n-1};d) = 1$. If $d = 1$, then an infinity of solution vectors of the Diophantine equation

$$D(x_1,\ldots,x_n;k_1,\ldots,k_{n-1};1) = (-1)^{v(n-1)}$$

is given by

$$(7.39)\qquad X_v = (A_0^{(v)}, A_0^{(v+1)},\ldots,A_0^{(v+n-1)}) \qquad (v=0,1,\ldots)$$

and $D(x_1,\ldots,x_n;k_1,\ldots,k_{n-1},1) = -1$ is solvable for v odd and n even. We have based the theory of solving $D(x_1,\ldots,x_n;k_1,\ldots,k_{n-1};d) = \pm 1$ on the condition that the polynomial $F(x) = x^n + k_1x^{n-1} + \cdots + k_{n-1} - d$, $d|k_i$ $(i=1,\ldots,n-1)$ be irreducible; the reader will verify easily that we can now drop this restriction as far as solving the above equation is concerned. We shall illustrate our theory with numeric examples.

Example 15. We obtain from (7.38), for n = 3,

$$D(x_1,x_2,x_3;k_1,k_2;d) =$$

$$\begin{vmatrix} x_1 & dx_3-k_2x_2 & dx_2 \\ x_2 & x_1+k_1x_2 & dx_3 \\ x_3 & x_2+k_1x_3 & x_1+k_1x_2+k_2x_3 \end{vmatrix}$$

$$= \begin{vmatrix} x_1, & dx_3-k_1x_1-k_2x_2, & dx_2 \\ x_2, & x_1, & dx_3 \\ x_3, & x_2, & x_1+k_1x_2+k_2x_3 \end{vmatrix}.$$

Let be $k_1 = 3$, $k_2 = -12$; $d = -3$. The equation $D(x_1,x_2,x_3;k_1,k_2,d) = 1$ takes the form

$$\begin{vmatrix} x_1, & -3x_3-3x_1+12x_2, & -3x_2 \\ x_2, & x_1, & -3x_3 \\ x_3, & x_2, & x_1+3x_2-12x_3 \end{vmatrix} = 1.$$

From the period

$$b^{(0)} = (3,\ -12)$$
$$b^{(1)} = (3,\ 4)$$
$$b^{(2)} = (-1,\ 4)$$

we calculate

$$A_0^{(3)} = 1; \quad A_0^{(4)} = 4; \quad A_0^{(5)} = 15;$$

$$A_0^{(6)} = -167; \quad A_0^{(7)} = -619; \quad A_0^{(8)} = -2294.$$

So

$$x_1 = (1, 4, 15)$$

$$x_2 = (-167, -619, -2294)$$

are solution vectors of $D(x_1,x_2,x_3;3,-12,3) = 1$.

Example 16. We obtain from (7.38), for $n = 4$,

$$D(x_1,\ldots,x_4;k_1,k_2,k_3;d) =$$

$$\begin{vmatrix} x_1, & dx_4-k_2x_2-k_3x_3, & dx_3-k_3x_2, & dx_2 \\ x_2, & x_1+k_1x_2, & dx_4-k_3x_3, & dx_3 \\ x_3, & x_2+k_1x_3, & x_1+k_1x_2+k_2x_3, & dx_4 \\ x_4, & x_3+k_1x_4, & x_2+k_1x_3+k_2x_4, & x_1+k_1x_2+k_2x_3+k_3x_4 \end{vmatrix}.$$

Let be $k_1 = 2$, $k_2 = 4$, $k_3 = 6$; $d = 2$. The equation $D(x_1,\ldots,x_4;\ 2,4,6;2) = 1$ takes the form

$$\begin{vmatrix} x_1, & 2x_4-2x_1-4x_2-6x_3, & 2x_3-6x_2, & 2x_2 \\ x_2, & x_1, & 2x_4-6x_3, & 2x_3 \\ x_3, & x_2, & x_1+2x_2+4x_3, & 2x_4 \\ x_4, & x_1, & x_2+2x_3+4x_4, & x_1+2x_2+4x_3+6x_4 \end{vmatrix} = 1.$$

From the period

$$b^{(0)} = (2, 4, 6),$$

$$b^{(1)} = (2, 4, 3),$$

$$b^{(2)} = (2, 2, 3),$$

$$b^{(3)} = (1, 2, 3),$$

we calculate

$$A_0^{(4)} = 1 \; ; \quad A_0^{(5)} = 3 \; ; \quad A_0^{(6)} = 11 \; ; \quad A_0^{(7)} = 40 \quad ,$$

so $X_1 = (1 \; , 3 \; , 11 \; , 40)$ is a solution of the above equation.

If we specify in (7.38)

(i) $k_1 = k_2 = \cdots = k_{n-1} = d = 1,$

(ii) $k_1 = k_2 = \cdots = k_{n-2} = 0,$

then the solutions vectors $X_v = (A_0^{(v)}, A_0^{(v+1)}, \ldots, A_0^{(v+n-1)})$ in the case (i) can be stated explicitly by means of formula (7.1), and the solution vectors $X_t = (A_0^{(tn)}, A_0^{(tn+1)}, \ldots, A_0^{(tn+n-1)})$ in the case (ii) — by means of formula (7.7).

We shall conclude this chapter with a special case of the Generalized Pell equation. Let ε be a unit in $Q(w)$, $w = \sqrt[n]{m}$ of the form

$$\varepsilon = x_1 + x_2 w; \quad x_3 = x_4 = \cdots = x_n = 0. \tag{7.40}$$

Formula (7.21) takes the form

$$N(\varepsilon) = \begin{vmatrix} x_1 & x_2 & 0 & 0 & & \dots & 0 & 0 \\ 0 & x_1 & x_2 & 0 & & \dots & 0 & 0 \\ 0 & 0 & x_1 & x_2 & 0 & \dots & 0 & 0 \\ - & - & - & - & - & - & - & - \\ mx_2 & 0 & 0 & & & \dots & 0 & x_1 \end{vmatrix},$$

so that

(7.41) $$x_1^{\,n} + (-1)^{n-1} m x_2^{\,n} = \pm 1.$$

For $n = 2$, (7.53) is the Pell equation $x_1^{\,2} - mx_2^{\,2} = \pm 1$, and for $n = 3$, $x_1^{\,3} + mx_2^{\,3} = \pm 1$. Thus (7.53) is not, as could have been thought, the Generalized Pell equation, but only a special case of it. Equation (7.53) has solutions, if and only if a unit ε in $Q(w)$ exists having the form (7.52). For $m = D^n \pm 1$, there is always one solution, since, as we know, $\varepsilon = -D + w$ in this case. The question whether there exist solutions of (7.53) for any $n \geqslant 3$ is undecided and seems very difficult; by Thue-Siegel-Roth's Theorems there is at most a finite number of solutions for a given n and a fixed m. The case $n = 3$, namely

(7.42) $$x^3 + my^3 = 1; \quad m \neq c^3, \ c \in I,$$

has been profoundly studied by B. Delaunay [10] and T. Nagell [20,1]. For the equation **(7.42)** Nagell has proved the following theorem: it has at most one (nontrivial, viz. different from $x = 1$, $y = 0$) solution; if x_1, y_1 is a solution, then

$$\varepsilon = x_1 + \sqrt[3]{m}\, y_1$$

is either a fundamental unit of $Q(\sqrt[3]{m})$ or its square; the latter can happen only for finitely many values of m.

Let m have the structure

$$(7.43)\quad \begin{cases} m = D^3 + k, \qquad k \neq \pm 1, \\ \pm k = d,\ d^2,\ dD,\ d^2D,\ 3d,\ 3d^2,\ 3dD,\ 3d^2D \\ \qquad d \mid D;\quad d, D \in N. \end{cases}$$

Then, as was shown in Chapter 6,

$$(7.44)\qquad \varepsilon = \frac{(w - D)^3}{k}, \qquad w = \sqrt[3]{m}$$

is a unit in $Q(w)$. By Stender's theorem, ε is fundamental for certain values of k; this author conjectures that ε is fundamental for all values of k from (7.43), with at most a finite number of exceptions for certain cases of numerical values of k. Now, from (7.44)

$$(7.45)\qquad \varepsilon = 1 + \frac{3D^2}{k} w - \frac{3D}{k} w^2$$

and (7.45) is not of the form $x + yw$. Squaring (7.45) we obtain

$$\varepsilon^2 = a + bw + \left(\frac{9D^4}{k^2} - \frac{6D}{k}\right) w^2;$$

demanding that the coefficient of w^2 vanish, we obtain

$$3D^3 = 2k;$$

it is easily verified that this equation has no solution for the values of k from (7.43) but one, viz. $k = 3dD$; we obtain, for this value of k,

$$3D^3 = 6dD, \qquad D^2 = 2d,$$

which has the only solution

$$D = d = 2;\quad k = 12;\quad m = 20.$$

From (7.45)' we obtain, for $D = 2$, $k = 12$

$$\varepsilon = 1 + w - \frac{1}{2} w^2, \qquad \varepsilon^2 = -19 + 7w,$$

and indeed $(-19, 7)$ is a solution of $x^3 + 20y^3 = 1$. This result was obtained by Nagell in a different and quite laborious way.

In [8] J. H. E. Cohn has given necessary conditions for the solubility of (7.54), investigating the structure of m modulo 9. In [2,p] this author has shown that for many infinite classes of m, these conditions are not sufficient. But the question, how to calculate a solution of (7.42) if one exists, is still open. The best way is probably to find, if any, a unit $\varepsilon = x + yw$ in $Q(w)$ by the JPA with the outer T-function of $a^{(0)} = (w, w^2)$. With a computer program, the author has found units of the form $x + yw$ for the following cases:

$m = 19$; $\varepsilon = -8 + 3w$;

$m = 20$; $\varepsilon = -19 + 7w$;

$m = 37$; $\varepsilon = 10 - 3w$;

$m = 43$; $\varepsilon = -7 + 2w$;

$m = 91$; $\varepsilon = 9 - 2w$;

$m = 254$; $\varepsilon = 19 - 3w$;

$m = 422$; $\varepsilon = -15 + 2w$;

$m = 614$; $\varepsilon = 17 - 2w$;

$m = 635$; $\varepsilon = 361 - 42w$

$m = 651$; $\varepsilon = -26 + 3w$;

$m = 813$; $\varepsilon = 28 - 3w$.

REFERENCES

/1/ . BACHMAN, P. : "Zur Theorie von Jacobi's Kettenbruch-Algorithmen", Jour.f.d.reine angew.Math., 75, 25-34, 1873.

/2/ . BERNSTEIN,L.: a) "The Modified Algorithm of Jacobi-Perron", Memoirs Amer.Math.Soc., 67, 1-44, 1966.

b) "Periodical Continued Fractions of Degree n by Jacobi's Algorithm", Jour.f.d.reine angew Math., 213, 31-38, 1964.

c) "New Infinite Classes of Periodic Jacobi-Perron Algorithms", Pacific Jour.Math., 16, 1-31, 1965.

d) "Einheitenberechnung in kubischen Koerpern mittels des Jacobi-Perronschen Algorithmus aus der Rechenanlage", J.f.d.reine angew Math., in print.

e) "Periodicity of Jacobi's Algorithm for a Special Type of Cubic Irrationals", J.f.d.reine angew.Math., 213, 134-146, 1964.

f) "Periodicity in Cubic Fields", Jour.Number Theory, in print.

g) "Representation of $(D^n - d)^{1/n}$ as a Periodic Continued Fraction by Jacobi's Algorithm", Math. Nachrichten, 1019, 179-200, 1965.

h) "Der B-Algorithmus und seine Anwendung", J.f.d.reine angew.Math., 227, 150-177, 1967.

i) "An Explicit Summation Formula and its Application", Proc.Amer.Math.Soc., 25, No.2.,323-334, 1970.

j) "A Probability Function for Partitions", Amer. Math.Monthly, 75, No.8, 882-886, 1968.

k) "The Linear Diophantine Equation in n Variables and its Application", The Fibonacci Quarterly, Special Issue, 1-52, (June) 1968.

l) "Generating Functions for Summation Formulas", Math.of Comput. (Amer.Math.Soc.), Submitted.

m) "The Generalized Pellian Equation", Transact. Amer.Math.Soc., 127, 76-89, 1967.

n) "Periodische Jacobi-Perronsche Algorithmen", Archiv d.Math., 15, 421-429, 1964.

p) "Infinite non-Solubility Classes of the Delaunay-Nagell Equation", Jour.London Math.Soc., **Vol.3 , Part 1 ,pp.118 - 120.**

/3/ . BERNSTEIN,L. and HASSE, H. : a) Einheitenberechnung mittels des Jacobi-Perronschen Algorithmus", J.f.d.reine angew Math., 218, 51-69, 1965.

b) "An Explicit Formula for the Units of an Algebraic Number Field of Degree n $\geqslant$ 2", Pacific Jour.Math., 30, (no.2) 293-365, 1969.

c) "Explicit Determination of the Matrices in Periodic Algorithms of the Jacobi-Perron Type with Application to Generalized Fibonacci Numbers with Time Impulses", Fibonacci Quarterly, 7 Special Issue (No.4) 394-436, 1969.

/4/ . BERWICK, W.E.H. : "Ideal Numbers that depend on a Cubic Irrationality", Proc.London Math.Soc., 12, 393-439, 1913.

/5/ . BILEVICH, K.K. : "On Units in Algebraic Fields of third and fourth Degree", (Russian) Math.Sbornik, 40 (82), 123-136, 1956.

/6/ . BISSINGER, B.H. : "A Generalization of Continued Fractions", Bull.Amer.Math.Soc., 50, 868-876, 1944.

/7/ . EVERETT, C.J. : "A Representation for Real Numbers", Bull.Amer.Math.Soc., 52, 861-869, 1946.

GOOD, J.I. : "The Fractional Dimensional Theory of Continued Fractions", Proc.Cambridge Philos. Soc., 37, 199-228, 1941.

LEIGHTON, W. : "Proper Continued Fractions", Amer.Math.Monthly, 47, 274-280, 1940.

PITCHER, T.S., KIMNEY, J.R. : "The Dimension of some Sets Defined in Terms of f - Expansions", Zeitschr.Wahrscheinl.verw.Geb., 4, 293-315, 1966.

RENYI, A. : "Representation for Real Numbers and their Ergodic Properties", Acta Math.Acad.Sci. Hungar., 8, 477-493, 1957.

ROGERS, C.A. : "Some Sets of Continued Fractions", Proc.London Math.Soc., (3) 14, 29-44, 1964.

/8/ . COHN, J.H.E. : "The Diophantine Equation $x^3 + dy^3 = 1$", Jour. London Math.Soc., 42, 750-752, 1967.

/9/ . DAUS, P.H. : "Normal Ternary Continued Fraction Expansions for the Cubic Roots of Integers", Amer.Jour.Math., 44, 29-296, 1922.

"Normal Ternary Continued Fraction Expansions for Cubic Irrationals", Amer.Jour.Math., 51, 67-93, 1929.

/10/ . DELONE, B.N. and FADDEV, D.K. : "The Theory of Irrationalities of the Third Degree", Translation of Math.Monographs, Amer. Math.Soc., 1964.

/11/ . DUNFORD, N. and MILLER, D.S. : "On the Ergodic Theorem", Transact.Amer.Math.Soc., 60, 538-549, 1946.

/12/ . FINKELSTEIN, R. and LONDON, H. : "On the Diophantine Equation $y^3 + p = x^2$", to appear.

/13/ . HARTMAN, S., MARCZEWSKI, E. and RYLL NARDZEWSKI, C. : "Theoremes ergodiques et leurs Application", Colloq.Math., 2, 109-123, 1950.

/14/ . HERMITE, CH. : "Letter to C.G.J. Jacobi", J.f.d.reine angew Math., 40, 286, 1839.

/15/ . JACOBI, C.G.J. : "Allgemeine Theorie der kettenbruchaehnlichen Algorithmen, in welchen jede Zahl aus drei vorhergehenden gebildet wird", J.f.d.reine angew.Math., 69, 29-64, 1869.

/16/ . KHINTCHINE, A.J. : "Kettenbrueche", Leipzig,Teubner.

/17/ . KNOPP, K. : "Mengentheoretische Behandlung einiger Probleme der Diophantischen Approximationen der transfiniten Wahrscheinlichkeiten", Math.Ann., 95, 409-426, 1926.

/18/ . LEHMER, D.N. : "On Jacobi's Extension of the Continued Fraction Algorithm", Proc.Nat.Acad.Sci.,U.S.A., 4, 360-364, 1918.

/19/ . LINNIK, Yu. : "Ergodic Properties of Algebraic Fields", Ergebnisse der Mathematik und ihrer Grenzgebiete, Band 45, Springer-Verlag New York, 1968.

/20/ . MINKOWSKI, H. : "Ein Kriterium fuer die algebraischen Zahlen", Goettinger Nachrichten, 1899.

"Ueber periodische Approximationen algebraischer Zahlen", Acta Math., § 2, 26.

/21/ . NAGELL, T. : "Solution Complete de quelque Equation Cubique a deux Indeterminess", Jour.Math.Pures et Appliques, 4, 290-270, 1925.

/22/ . PERRON, O. : "Grundlagen fuer eine Theorie des Jacobischen Kettenbruchalgorithmus", Math.Ann., 64, 1-76, 1907.

/23/ . RAAB, J.A. : "Some non-Jacobian Ternary Continued Fractions", Dissertation, Wisconsin State Univer. Oshkosh, 1968.

/24/ . RANEY, G. : "Generalization of the Fibonacci Sequence to n Dimensions", Canadian Jour.Math., 332-349, 1967.

/25/ . SCHWEIGER, F. : "Geometrische und elementar metrische Saetze ueber den Jacobischen Algorithmus", Sitzungsber.Oesterr.Akad.Wiss.Math.=naturwiss.Klasse, Abt.II, 173, 42-59, 1964.

"Metrische Saetze ueber den Jacobischen Algorithmus", Monatshefte f.Math., 69, 243-255, 1965.

"Ergodische Theorie des Jacobischen Algorithmus", Acta Arith., 11, 451-460, 1966.

"Existenz eines invarianten Masses beim Jacobischen Algorithmus", Acta Arith., XII, 263-268, 1967.

"Induzierte Masse und Jacobischer Algorithmus", Acta Arith., XIII, 419-422, 1968.

"Mischungseigenschaften und Entropie beim Jacobischen Algorithmus", J.f.d.reine angew. Math., 229, 50-56, 1968.

/26/ . STENDER, J. : "Ueber die Grundeinheit fuer spezielle unendliche Klassen reiner kubischer Zahlkoerper", Abhandlungen Math.Seminar Univers.Hamburg., 33, 203-215, 1969.

/27/ . WORONOJ, G.F. : "On a Generalization of the Algorithm for Continued Fractions", (Russian) Warsaw 1896, Dissertation, unpublished.